Cross-Cutting Themes for U.S. Contributions to the UN Ocean Decade

T0003577

Committee on Cross-Cutting Themes for U.S. Contributions to the Ocean Decade

Ocean Studies Board

Division on Earth and Life Studies

A Consensus Study Report of

The National Academies of
SCIENCES · ENGINEERING · MEDICINE

THE NATIONAL ACADEMIES PRESS
Washington, DC
www.nap.edu

THE NATIONAL ACADEMIES PRESS 500 Fifth Street, NW Washington, DC 20001

This activity was supported by a contract between the National Academy of Sciences and the National Aeronautics and Space Administration under Award Number 10002633. Any opinions, findings, conclusions, or recommendations expressed in this publication do not necessarily reflect the views of any organization or agency that provided support for the project.

International Standard Book Number-13: 978-0-309-27311-4
International Standard Book Number-10: 0-309-27311-0
Digital Object Identifier: https://doi.org/10.17226/26363

This publication is available from the National Academies Press, 500 Fifth Street, NW, Keck 360, Washington, DC 20001; (800) 624-6242 or (202) 334-3313; http://www.nap.edu.

Copyright 2022 by the National Academy of Sciences. All rights reserved.

Printed in the United States of America

Cover credit: Paul Johnson of the University of New Hampshire for the globe image.

Suggested citation: National Academies of Sciences, Engineering, and Medicine. 2022. *Cross-Cutting Themes for U.S. Contributions to the UN Ocean Decade*. Washington, DC: The National Academies Press. https://doi.org/10.17226/26363.

The National Academies of

SCIENCES • ENGINEERING • MEDICINE

The **National Academy of Sciences** was established in 1863 by an Act of Congress, signed by President Lincoln, as a private, nongovernmental institution to advise the nation on issues related to science and technology. Members are elected by their peers for outstanding contributions to research. Dr. Marcia McNutt is president.

The **National Academy of Engineering** was established in 1964 under the charter of the National Academy of Sciences to bring the practices of engineering to advising the nation. Members are elected by their peers for extraordinary contributions to engineering. Dr. John L. Anderson is president.

The **National Academy of Medicine** (formerly the Institute of Medicine) was established in 1970 under the charter of the National Academy of Sciences to advise the nation on medical and health issues. Members are elected by their peers for distinguished contributions to medicine and health. Dr. Victor J. Dzau is president.

The three Academies work together as the **National Academies of Sciences, Engineering, and Medicine** to provide independent, objective analysis and advice to the nation and conduct other activities to solve complex problems and inform public policy decisions. The National Academies also encourage education and research, recognize outstanding contributions to knowledge, and increase public understanding in matters of science, engineering, and medicine.

Learn more about the National Academies of Sciences, Engineering, and Medicine at **www.nationalacademies.org.**

The National Academies of

SCIENCES · ENGINEERING · MEDICINE

Consensus Study Reports published by the National Academies of Sciences, Engineering, and Medicine document the evidence-based consensus on the study's statement of task by an authoring committee of experts. Reports typically include findings, conclusions, and recommendations based on information gathered by the committee and the committee's deliberations. Each report has been subjected to a rigorous and independent peer-review process and it represents the position of the National Academies on the statement of task.

Proceedings published by the National Academies of Sciences, Engineering, and Medicine chronicle the presentations and discussions at a workshop, symposium, or other event convened by the National Academies. The statements and opinions contained in proceedings are those of the participants and are not endorsed by other participants, the planning committee, or the National Academies.

For information about other products and activities of the National Academies, please visit www.nationalacademies.org/about/whatwedo.

COMMITTEE ON CROSS-CUTTING THEMES FOR U.S. CONTRIBUTIONS TO THE OCEAN DECADE

LARRY A. MAYER (NAE), *Chair*, University of New Hampshire, Durham
MARK R. ABBOTT, *Vice Chair,* Woods Hole Oceanographic Institution, Massachusetts
CAROL ARNOSTI, University of North Carolina at Chapel Hill
CLAUDIA BENITEZ-NELSON, University of South Carolina, Columbia
ANJALI BOYD, Duke University, Durham, North Carolina
ANNIE BRETT, Levin College of Law, University of Florida, Gainesville
THOMAS S. CHANCE, ASV Global, LLC (ret.), Broussard, Louisiana
DANIEL COSTA, University of California, Santa Cruz
JOHN R. DELANEY, University of Washington (ret.), Seattle
ANGEE DOERR, Oregon State University, Oregon Sea Grant, Corvallis
SCOTT GLENN, Rutgers University, New Brunswick, New Jersey
PATRICK HEIMBACH, The University of Texas at Austin
MARCIA ISAKSON, The University of Texas at Austin
LEKELIA JENKINS, Arizona State University, Tempe
SANDRA KNIGHT, University of Maryland, College Park
NANCY KNOWLTON (NAS), Smithsonian Institution (ret.), Washington, District of Columbia
ANTHONY MacDONALD, Monmouth University, West Long Branch, New Jersey
JACQUELINE McGLADE, University College London, United Kingdom
THOMAS J. MILLER, University of Maryland, Solomons
S. BRADLEY MORAN, University of Alaska Fairbanks
RUTH M. PERRY, Shell Exploration & Production Company, Houston, Texas
JAMES SANCHIRICO, University of California, Davis
MARK J. SPALDING, The Ocean Foundation, Washington, District of Columbia
LYNNE D. TALLEY, Scripps Oceanography, San Diego, California
ROBERT S. WINOKUR, Michigan Tech Research Institute, Silver Spring, Maryland
GRACE C. YOUNG, X, Alphabet's Moonshot Factory, Mountain View, California

Staff

SUSAN ROBERTS, Director
VANESSA CONSTANT, Associate Program Officer
SHELLY-ANN FREELAND, Financial Business Partner (through January 2022)
THANH NGUYEN, Financial Business Partner
ELIZABETH COSTA, Senior Program Assistant

OCEAN STUDIES BOARD

CLAUDIA BENITEZ-NELSON, *Chair*, University of South Carolina, Columbia
MARK R. ABBOTT, Woods Hole Oceanographic Institution, Massachusetts
CAROL ARNOSTI, University of North Carolina at Chapel Hill
LISA M. CAMPBELL, Duke University, Durham, North Carolina
THOMAS S. CHANCE, ASV Global, LLC (ret.), Broussard, Louisiana
DANIEL COSTA, University of California, Santa Cruz
JOHN R. DELANEY, University of Washington (ret.), Seattle
SCOTT GLENN, Rutgers University, New Brunswick, New Jersey
PATRICK HEIMBACH, The University of Texas at Austin
MARCIA ISAKSON, The University of Texas at Austin
LEKELIA JENKINS, Arizona State University, Tempe
NANCY KNOWLTON (NAS), Smithsonian Institution (ret.), Washington, District of Columbia
ANTHONY MacDONALD, Monmouth University, West Long Branch, New Jersey
THOMAS J. MILLER, University of Maryland, Solomons
S. BRADLEY MORAN, University of Alaska Fairbanks
RUTH M. PERRY, Shell Exploration & Production Company, Houston, Texas
JAMES SANCHIRICO, University of California, Davis
MARK J. SPALDING, The Ocean Foundation, Washington, District of Columbia
ROBERT S. WINOKUR, Michigan Tech Research Institute, Silver Spring, Maryland

Staff

SUSAN ROBERTS, Director
STACEE KARRAS, Senior Program Officer
KELLY OSKVIG, Senior Program Officer
EMILY TWIGG, Senior Program Officer
ALEXANDRA SKRIVANEK, Associate Program Officer (through February 2022)
VANESSA CONSTANT, Associate Program Officer
SHELLY-ANN FREELAND, Financial Business Partner (through January 2022)
THANH NGUYEN, Financial Business Partner
KENZA SIDI-ALI-CHERIF, Senior Program Assistant (through March 2022)
ELIZABETH COSTA, Senior Program Assistant
GRACE CALLAHAN, Program Assistant

Preface

The United Nations (UN) Decade of Ocean Science for Sustainable Development 2021–2030 (UN Ocean Decade) provides a pathway to support and implement a globally coordinated scientific effort to develop solutions for ensuring the health and sustainability of our life-sustaining oceans. For many who have devoted their lives to the study of the ocean, the UN Ocean Decade is an opportunity to highlight the critical importance of the oceans to the long-term well-being of humankind. The UN Ocean Decade's Implementation Plan (which benefited tremendously from the input of U.S. Executive Planning Group representatives—Christa von Hillebrandt-Andrade, Margaret Leinen, Craig McLean, and Linwood Pendleton) describes outcomes that, if achieved, would ensure that we have "the ocean we need for the future we want." The challenge to the global community is to define and execute the science needed to make these outcomes a reality.

While the UN Ocean Decade provides overarching guidance, much of the work and particularly the funding for UN Ocean Decade efforts will happen on a national or regional level. Recognizing the fundamental importance of healthy and sustainable oceans to the United States and the historic leadership role that the United States has played in ocean sciences, Craig McLean, Assistant Administrator for Oceanic and Atmospheric Research and Acting Chief Scientist at the National Oceanic and Atmospheric Administration, led the effort to promote UN Ocean Decade activities both within the federal government and through the establishment of the nongovernmental, U.S. National Committee for the UN Decade of Ocean Science for Sustainable Development (U.S. National Committee), hosted by the National Academies of Sciences, Engineering, and Medicine.

The U.S. National Committee and the nation owe Craig a tremendous debt for his inspirational and tireless efforts to support the concepts and principles of the UN Ocean Decade and particularly the role that the United States can play in it. His challenge to the U.S. National Committee—to inspire the U.S. community to develop audacious and transformative science in support of the UN Ocean Decade—led then U.S. National Committee vice-chair Rick Spinrad to propose the concept of "Ocean-Shots." The tremendous response to the call for Ocean-Shots clearly demonstrated the creativity, breadth, and capacity of the U.S. community and set the framework for this report.

In October 2021, the White House Office of Science and Technology Policy, with funding from the National Aeronautics and Space Administration, called on the National Academies to produce this consensus report and recommend several cross-cutting themes based on the Ocean-Shots and complemented by U.S. ocean science priorities and UN Ocean Decade outcomes. This led to frenetic activity on the part of the consensus committee, the Committee on Cross-Cutting Themes for U.S. Contributions to the Ocean Decade, which was made more difficult by the inability to meet in person. The committee's hard work and thoughtful input helped us meet very short deadlines. Most importantly, credit must be given to Ocean Studies Board director Susan Roberts and associate program officer Vanessa Constant who, with the excellent assistance of senior program assistant Elizabeth Costa, worked tirelessly to put together a coherent report.

The recommendations of this report offer what we believe to be a bold plan for the future of ocean science in support of sustainable development. The suggested themes offer frameworks for developing solutions to critical issues of ocean health and global well-being, while also defining new and more inclusive ways of doing ocean science. At this point, however, they are only frameworks. Whether they grow into fully defined research programs that will lead to the "ocean we need for the future we want" will depend on the response of those to whom this report is submitted. With leadership from the federal agencies, in conjunction with international efforts and contributions from other sectors, the research community will provide the science necessary to achieve a sustainable ocean.

Larry A. Mayer, *Chair*
Committee on Cross-Cutting Themes for
U.S. Contributions to the Ocean Decade

Acknowledgments

The committee would like to thank Tom Drake, Jane Lubchenco, Craig McLean, and Terry Quinn for their presentations at the November 12, 2021, meeting of the U.S. National Committee for the United Nations Decade of Ocean Science for Sustainable Development. In addition, the committee would like to acknowledge the thoughtful responses received during the posting of the draft themes for public comment. These contributions informed the committee's revisions of the themes as presented in the report.

Reviewers

This Consensus Study Report was reviewed as a draft by individuals chosen for their diverse perspectives and technical expertise. The purpose of this independent review is to provide candid and critical comments that will assist the National Academies of Sciences, Engineering, and Medicine in making each published report as sound as possible and to ensure that it meets the institutional standards for quality, objectivity, evidence, and responsiveness to the study charge. The review comments and draft manuscript remain confidential to protect the integrity of the deliberative process.

We thank the following individuals for their review of this report:

JULIE BRINGHAM-GRETTE, University of Massachusetts Amherst
MELBOURNE BRISCOE, OceanGeeks, LLC
KEITH CRIDDLE, University of Alaska Fairbanks
TIMOTHY GALLAUDET, Ocean STL Consulting
CASSIA GALVAO, Texas A&M at Galveston
ASHANTI JOHNSON, MatlScience, Inc.
GERALDINE KNATZ (NAE), University of Southern California, Los Angeles
MARGARET LEINEN, Scripps Institution of Oceanography
MOLLY McCAMMON, Alaska Ocean Observing System
NANCY RABALAIS (NAS), Louisiana State University

Although the reviewers listed above provided many constructive comments and suggestions, they were not asked to endorse the conclusions or recommendations of this report nor did they see the final draft before its release. The review of this report was overseen by **ROBERT DUCE,** Texas A&M University, and **BONNIE McCAY,** Rutgers University (ret.). They were responsible for making certain that an independent examination of this report was carried out in accordance with the standards of the National Academies and that all review comments were carefully considered. Responsibility for the final content rests entirely with the authoring committee and the National Academies.

Contents

Summary

The United Nations (UN) Decade of Ocean Science for Sustainable Development 2021–2030 (UN Ocean Decade) offers a rare opportunity to bring the global community together to design and implement the science needed to ensure the health and sustainability of the ocean. The UN Ocean Decade further raises international recognition of the critical importance of the ocean and ocean science to the long-term well-being of humankind. The UN effort aspires to provide "the science we need for the ocean we want" by the conclusion of the UN Ocean Decade. As part of this process, the United Nations, through the Intergovernmental Oceanographic Commission, is creating an organizational structure to support and manage UN Ocean Decade activities. This includes a call for the establishment of national committees designed to act as an interface between national efforts and the UN Ocean Decade and to provide a focal point for country-level coordination of UN Ocean Decade–related activities. The United States, through a request from the National Oceanic and Atmospheric Administration, has responded with the establishment of the U.S. National Committee for the UN Decade of Ocean Science for Sustainable Development (U.S. National Committee), hosted by the National Academies of Sciences, Engineering, and Medicine (the National Academies). The U.S. National Committee operates independently of the federal agencies but maintains regular communications to ensure coordination on efforts in support of UN Ocean Decade activities.

The U.S. National Committee seeks to inspire, foster, and, where possible, facilitate the organization of U.S. contributions to the UN Ocean Decade–related ocean science activities with the intent that they will lead to funded programs supporting UN Ocean Decade objectives and advancing ocean science in the United States. Thus, a critical role of both the UN Ocean Decade and the U.S. National

Committee is to facilitate and encourage sponsorship of research to help meet the scientific challenges of the UN Ocean Decade. The UN Ocean Decade aims to advance progress toward the United Nations' Sustainable Development Goals by raising awareness of opportunities in ocean sciences and enabling the connection between scientists with promising research concepts and potential funders in the private and public sectors. To foster collaboration across sectors, the U.S. National Committee has established a communications and information hub through the creation of a dedicated website, public meetings, a Nexus consisting of U.S. organizations with an interest in the UN Ocean Decade, newsletters, and social media.

In the fall of 2020, the U.S. National Committee held a public virtual meeting to issue a call for "Ocean-Shots," defined as ambitious, transformational research concepts that draw inspiration and expertise from multiple disciplines and fundamentally advance ocean science for sustainable development. Information about the Ocean-Shot request was shared on the U.S. National Committee website,[1] via the U.S. National Committee newsletter, and through announcements shared by federal agencies and nonfederal organizations. The U.S. National Committee's vision was to foster the creation of programs of such broad-reaching scale and relevance that they would capture the imagination of the science community, funders, the public, and policy makers, much like the "moonshot" mission to the Moon of the 1960s.[2]

The call for Ocean-Shots offered a series of criteria for the submissions. These included demonstrating potential transformative impact; engaging scientific, technical, and other sectors outside the traditional ocean sciences; building capacity and a cadre of next-generation ocean scientists; and directly addressing at least 1 of the 10 "Challenges" set forth in the UN Ocean Decade's Implementation Plan (see Table 1.1). The U.S. National Committee anticipated that many of the ideas submitted as Ocean-Shots could be developed into projects or programs for submission to the UN Ocean Decade for endorsement. Consequently, the criteria also included identification of opportunities for international participation and collaboration. The Ocean-Shots were not meant to duplicate the UN Ocean Decade call for actions; rather, these submissions were meant to provide the impetus for the development of new projects and programs over the course of the decade, including building on UN Ocean Decade–endorsed actions with significant input from the U.S. ocean community. The response to the call for Ocean-Shots yielded more than 100 submissions, demonstrating the breadth, creativity, and enthusiasm of the U.S. ocean science and engineering communities.

Building on the robust response to the call for Ocean-Shots, the U.S. National Committee continues to foster further development of the Ocean-Shot concepts through coordination with the National Science and Technology Council's

[1] See nationalacademies.org/oceandecadeus.

[2] See https://www.jfklibrary.org/visit-museum/exhibits/past-exhibits/moon-shot-jfk-and-space-exploration.

Subcommittee on Ocean Science and Technology (SOST), representing the primary federal agencies that fund ocean science in the United States. In October 2021, the SOST asked the U.S. National Committee to identify between three and five cross-cutting themes emerging from the Ocean-Shots that complement U.S. ocean priorities, as described in *Science and Technology for America's Oceans: A Decadal Vision* (U.S. Decadal Vision; SOST, 2018); align with the overall goal of the UN Ocean Decade; and could potentially lead to new UN Ocean Decade–endorsed actions. The SOST further emphasized that these themes should bring a multidisciplinary approach to achieve "ocean science for sustainable development." With funding from the National Aeronautics and Space Administration, the Committee on Cross-Cutting Themes for U.S. Contributions to the Ocean Decade was appointed in October 2021 by the National Academies to undertake this task based on the membership of the U.S. National Committee and including four early career liaisons. This consensus study committee has proposed two foundational themes and four topical themes. The themes described here present frameworks for future development of research programs that provide sustainable solutions to the ocean challenges identified by the UN Ocean Decade. The foundational themes address the underlying infrastructure of how to conduct the science, ensuring that information and resources are accessible, applicable, and meet the needs of diverse communities through engagement in the development of research priorities and co-production of knowledge. The foundational themes will provide protocols and best practices for application within each of the four topical themes. The topical themes address fundamental aspects of sustainability and offer frameworks under which exciting new research programs can be further refined and developed. The successful realization of these themes will see the tangible demonstration of a healthier and more productive ocean and the implementation of practices to ensure the sustainability and continued improvement of the ocean as a whole.

FOUNDATIONAL THEMES

In identifying research themes for the U.S. contribution to the UN Ocean Decade, the committee recognized that there were two fundamental issues that needed to be addressed to ensure the success of all UN Ocean Decade research efforts. An Inclusive and Equitable Ocean, the first foundational theme, recognizes that increasing awareness, understanding, and access to the ocean can only be achieved through the integral involvement and support of a diverse, representative, and inclusive ocean community. Innovation is driven by diverse teams that facilitate novel thinking and adaptability.

Equity, inclusiveness, respect, fairness, and scientific integrity are core principles of the UN Ocean Decade that must be infused into all activities so that, in the words of the UN Ocean Decade, "no one is left behind" and all are empowered to contribute to the ocean science enterprise. **For this theme, success will**

be in the creation of an environment where different ideas and a diversity of viewpoints are sought and considered as integral to the process for identifying, developing, conducting, and applying ocean science. Scientific decisions will be made with interdisciplinary contributions, representing participation of many groups and recognizing the value of local knowledge and different knowledge systems. In addition, those who conduct ocean-related activities will reflect the diverse communities that use and benefit from ocean studies. It represents an opportunity to widen participation in the ocean enterprise to include groups not generally represented in key decision points for the design and use of novel ocean research. The need to expand participation crosses all aspects of ocean studies including the natural and social sciences; ocean literacy, education, and professional development; ocean governance, policy, and management; ocean industry; and others. In the United States, UN Ocean Decade activities will be distinguished by the level of engagement of communities outside the realm of traditional academic research, including the following:

- All four topical themes actively involve the broad community in the development of their research plans, as well as create, implement, and adaptively monitor the outcomes of an equity and inclusion action plan.
- All communities feel welcome to engage in Ocean Decade activities, with recognition of their interests, concerns, and active participation.
- Metrics are developed, applied, and documented such that each community can track meaningful points of progress in creating an equitable and inclusive Decade of Ocean Science for Sustainable Development.
- Ocean science, knowledge, educational materials, and social engagement are made accessible and broadly available through an open access platform.

Initial efforts on this foundational theme will define mechanisms for an inclusive and equitable ocean, establishing a framework and methodology to incorporate core principles into each topical theme. This would include early engagement and co-development of research topics and sustained community engagement in data collection, analysis, interpretation, and development of sustainable solutions. The initial outcome will be to elevate and prioritize equity and inclusion and incorporate these core principles during each topical theme's development. This is fundamental for the UN Ocean Decade and helps define and guide implementation of inclusive and equitable ocean science practices to achieve a more sustainable ocean for everyone.

An Ocean of Data, the second foundational theme, recognizes that there has been a major shift toward the principle of open access for data collected with government funds. However, open access only addresses one aspect of ensuring that the full value of data is realized. Because of the enormous range of data types, quality, and formats, and the enormous range of institutions and individuals who hold data, availability and usability are often not optimal and fail to consider the potential users of the data and their needs. Some ocean data from past research efforts remain locked away, unorganized, or still difficult to obtain from government agencies, companies, resource users, or researchers. The many examples of siloed data make it difficult to combine the insights and data across programs that are needed for a more interdisciplinary and holistic understanding of the interaction between societal interests and ocean conditions and trends. Such a democratization of data will be required to unleash the full potential of artificial intelligence (AI) and machine learning (ML).[3] The vision of adaptive and dynamic management approaches to our ocean will require open and accessible data and services that are driven by a user-centric and decentralized process. The scientific community (academic, government, and private) needs to rethink the intersections among observing systems, information systems, and knowledge services that are focused on delivering services and solutions, not just more data. This should include the development of strategies to relate place-based and qualitative knowledge to the more quantitative and large-scale, big data approaches and ways to incorporate many types of data, with consideration for different types of evidence that are important for the co-creation of knowledge.

An Ocean of Data will improve data availability and access through the development of a framework for implementing findable, accessible, interoperable, and reusable (FAIR) and collective benefit, authority to control, responsibility, and ethics (CARE) data principles and the creation of a path toward a digital ecosystem that delivers ubiquitous compute-intensive data services. Consideration of these digital ocean outcomes will also assist in the operationalization of each topical theme. By the end of the UN Ocean Decade, the committee envisions a network that connects the unconnected, truly engaging those who need knowledge services for ensuring a healthy ocean with those who can create those services. This will create a new era of scalability and flexibility, enabling new solutions to be developed and deployed rapidly, rather than relying on the static and inflexible systems of today that are designed primarily to address specific research questions. In partnership with existing UN Ocean Decade actions and private-sector initiatives, the United States could play a key role in the creation of an open, actionable, and equitable digital ecosystem for ocean knowledge across all ocean arenas.

[3] The potential for AI and ML applications in weather, climate, and Earth system science has been laid out in a number of recent reviews (e.g., Boukabara et al., 2020; Reichstein et al., 2019; and the monograph by Camps-Valls et al., 2021). Broadly, AI/ML tools have the potential to advance ocean sciences in the context of recognition and classification, anomaly detection, regression or inference, space- and/or time-dependent state prediction, acceleration of simulations, autonomous systems and active sampling, and emulation for uncertainty quantification. See NASEM (2022d).

TOPICAL THEMES

Based on the submitted Ocean-Shots, the committee identified four topical themes that represent promising areas for additional research investments consistent with the UN Ocean Decade's challenges and outcomes as well as identified priorities of the U.S. Decadal Vision. These topical themes represent opportunities for multidisciplinary collaboration that engage a broad swath of the ocean sciences, as well as disciplines outside the ocean sciences that could contribute new perspectives, technology, and education and engagement strategies. The themes are not meant to encompass every important topic of ocean research but rather to represent a subset of promising areas derived from the Ocean-Shot submissions, relevant to the goals of the UN Ocean Decade and U.S. ocean priorities, and selected to inspire additional investment in ocean science in support of sustainable development. **A key to the success of the topical themes will be the incorporation of the findings from the foundational themes to ensure that the research efforts involve diverse communities and perspectives and provide universal access to data and information that will define the way that science is practiced both during the UN Ocean Decade and beyond.**

The Ocean Revealed, the first topical theme, calls for the coordinated development and deployment of new techniques, technologies, and approaches, including citizen/community science where possible, to measure a range of ocean variables (biological, chemical, and physical) in understudied ocean regions to provide the critical understanding of ocean processes needed for sustainable development. For example, it builds on the potential of leveraging new technologies and existing infrastructure to develop an enhanced understanding of the ocean's acoustic environment and the rapid development of eDNA to enable characterization and monitoring of marine ecosystems. Processes that could be studied with these and other emerging technologies include heat, mass, and tracer transport; circulation and mixing; ecosystem health; and marine organism biodiversity and behavior. **By 2030, The Ocean Revealed would provide the observations necessary to support the Sustainable Development Goals and growth of the blue economy. As part of this initiative, observation needs will be identified through community engagement. This theme would provide globally coordinated observations of key ocean variables through the combination of newly developed inexpensive sensors, eDNA, and other "omics" techniques that can be broadly deployed; strategically located moorings measuring multiple parameters; and enhanced autonomous sensing capabilities aided by new developments in AI/ML and high-bandwidth satellite communications, all supported by a global array of active and passive acoustic sensors that take advantage of existing cabled infrastructure to provide power, communications, and positioning.** Such an observation "system of systems" would enable the following:

- More accurate forecasting and early warning of storm surge and tsunamis;
- Enhanced climate services and weather forecasts, including prediction, mitigation, and adaptation of global sea level, marine heatwaves, and other climatic events;
- Long-term monitoring of ocean hydrographic structure and sensitivity of marine organisms to change;
- Conservation of biodiversity through detection and monitoring of species; and
- Ecosystem-based fisheries management through enhanced monitoring of ecosystem components in addition to the target species.

The Restored and Sustainable Ocean, the second topical theme, takes a whole-ocean approach to intersections among the many Ocean-Shots focused on the restoration and sustainability of economically important ecosystems such as coral reefs, mangroves, and other coastal habitats, with an emphasis on nature-based solutions (NBSs). The challenges include the production of commercial fisheries and the development of mariculture that supports the livelihoods of coastal communities without loss of other critical ecosystem services. It further focuses on meeting the growing demands for forage fish in aquaculture and livestock operations while balancing the importance of their role in ecosystem health. Finally, emphasis is placed on characterizing the diversity and dynamics of underexplored ecosystems of the mesopelagic zone and the deep sea. This recognizes the need to understand the connectivity of the deep sea with mesopelagic and surface ecosystems and to characterize deep-seabed ecosystems in areas targeted for deep-sea mining as well as estimate recovery rates from disturbance by proposed seafloor mining activities. In all cases attention will be given to research that incorporates the human ecosystem into new strategies and approaches for sustaining ocean health and resources while supporting marine transportation, renewable energy, and evolving uses of the ocean. **By the end of the decade, co-production of social-ecological system knowledge will have yielded place-based approaches to more effectively restore and protect vulnerable coastal ecosystems and thus ensure provision of ecosystems valued in coastal regions. Scientific progress will support the implementation of evidence-based practices to restore and protect marine habitats with documented progress by the end of the decade. These practices will recognize and empower Indigenous voices and learn from and support traditional knowledge of coastal ecosystems for the benefit of all.** Included in the vision are the following:

- The transdisciplinary science necessary to develop a fuller understanding of the social-ecological system that supports fisheries and ocean aquaculture;

- The development of robust, climate-proof, ecosystem-based approaches that will be routinely used in commercial and recreational fisheries management;
- A dramatically improved understanding of deep-sea and seabed ecosystems globally, with an initial focus on areas targeted for ocean energy development, and to address a rapidly growing interest in mineral exploitation in the face of substantial challenges of conducting comprehensive baseline assessments of seafloor ecosystems; and
- A thriving ocean aquaculture industry that will contribute significantly to the economies of coastal communities and provide ecologically sustainable yields of marine consumables including seaweeds, shellfish, and finfish.

Ocean Solutions for Climate Resilience, the third topical theme, is focused on two key components of the extensive intersection between the ocean and climate. The first is the urgent need to anticipate and plan for coastal change in response to sea level rise and stronger, slower moving, and more frequent coastal storms. The second is the development and implementation of climate mitigation strategies including the expansion of renewable energy generation and development of carbon dioxide removal (CDR) and sequestration approaches.

For the first component, environmentally sustainable, equitable, and just adaptation strategies for responding to rising seas and increased coastal flooding will be developed with the affected coastal communities. This will require improvements in regional projections for sea level rise and coastal flooding, increased understanding of the capacity for NBSs to increase coastal resilience under a variety of climate scenarios, and co-development of knowledge and solutions for coastal communities, particularly those at high risk and with low capacity for adaptation. By applying social and human behavioral science studies of risk communication, researchers will be better able to work with coastal communities and Indigenous nations to collectively identify population vulnerabilities and information needs. Through this cooperative approach, improved predictions of weather and climate will be more integrated into community planning for, and response to, more frequent extreme events like flooding, heat, or drought. **By 2030, the social, cultural, and institutional drivers needed for a climate resilient future will be fully integrated in the science data and tools to enable an adaptable and resilient coastal zone.** This transdisciplinary approach will do the following:

- Allow for a common frame of reference for the development of flood/inundation maps that meet the variety of needs of state and federal agencies;
- Inform coastal planning that balances the adaptation of the communities and built systems disrupted by rising seas and extreme events with the economic drivers that encourage coastal development;

- Enable engagement of highly vulnerable U.S. coastal communities in the development of mitigation and adaptation strategies tailored to their particular needs;
- Improve coastal flooding projections to include compound and multi-hazard risks (e.g., storm surge and heavy rain) and the geomorphic transformation of the coastal zone under a range of climate change scenarios;
- Identify opportunities and challenges for ocean renewable energy; and
- Implement co-development of knowledge and solutions for coastal communities, particularly those at high risk and low capacity for adaptation.

For the second component, programs will focus on enhancing the ocean's natural ability for carbon capture through approaches such as ecosystem restoration; natural carbon capture technologies (e.g., coastal blue carbon); and habitat restoration for the recovery of depleted and endangered species subject to environmental stressors, including climate change. In addition, this theme highlights the potential for research into ocean-based renewable energy production and comprehensive analyses of material flows, such as critical minerals used in renewable technologies. **By 2030, science will have advanced to understand the merits, efficacy, limitations, and environmental risks of ocean-based CDR, including enhancing the ocean's natural uptake of carbon dioxide under a range of future climate scenarios. The research will provide a more complete understanding of the ethical, legal, and social contexts for ocean CDR and sequestration approaches. Reliable ocean renewable energy will provide a substantial fraction of the nation's electricity demand while balancing the interests of other critical ocean values (e.g., marine conservation, shipping, fisheries).**

Healthy Urban Seas, the fourth topical theme, takes a place-based approach with a focus on densely populated coastal centers where commerce, energy production, recreation, and fisheries depend on the nexus of the ocean environment with human activities. The concentration of human activities in and around these semi-enclosed waters provides a "test-bed" for investigating impacts, such as the run-off of plastics, nutrients, and other pollutants, and developing solutions for the health and resilience of urban seas and, by extension, the coastal ocean more generally. There is much opportunity to expand on the extensive science and data acquisition currently under way in our estuaries and coastal watersheds such as long-term studies of the Chesapeake Bay, Puget Sound, and Mississippi River Delta. Furthermore, ports are often a critical hub for these urban seas that offer potential partners for seeking low- or no emission solutions and protecting ecosystems. Ships and other vessels may provide ready-to-deploy monitoring platforms. Potential components of this theme include the development of long-term observing systems that include fixed infrastructure and autonomous systems programmed to monitor or track specific features (building on the output of The Ocean Revealed), the development of better methodologies and design guidance for resilient urban infrastructure, studies of the response of the coastal ecosystem and key species to multiple stressors, and efficacy of remediation efforts.

Urban seas provide a promising test for the development and application of digital twin models that encompass the physical, chemical, biological, and societal characteristics of the system. Such models are ideally suited for the relatively constrained yet complex components of an urban sea setting. The infrastructure and approaches established to monitor and understand U.S. urban seas could be used to model other systems around the world and form the basis for global collaborative projects, and have the potential to develop into a UN Ocean Decade–endorsed program. Importantly, these large population centers offer many opportunities for community engagement in problems that are directly relevant to their well-being. Ocean-related issues could readily be brought to the attention of policy makers with the involvement of diverse communities, provide enrichment activities and field trips for inclusion in school curricula, and enhance engagement through citizen/community science and public events. By integrating across disciplines and interests, Healthy Urban Seas would provide opportunities for creative development and sharing of ideas among a range of backgrounds, expertise, age groups, socioeconomic statuses, and public and private sectors: a pathway toward a more sustainable urban ocean. **By 2030, the urban sea will be well characterized through observation and modeling efforts leading to clean waters, enhanced coastal resilience, and greater community engagement.** This will include the following:

- Infrastructure to support the blue economy;
- Publicly available quantification of the flux of important properties, such as pollutant concentrations, from one water body to another;
- Major stakeholders identified and engaged in monitoring, research, education, and management applications;
- An urban sea defined and understood through observational and modeling efforts, such as the establishment of a digital twin;
- Measurable improvements in urban sea health through mitigation of pollution sources; and
- Development of a community culture that understands the importance of, and plays an active role in, ensuring a safe and healthy ocean through community engagement (ocean cultural literacy).

Looking across all six themes, several common topics emerge. This is not unexpected; overlapping topics illustrate the interconnections among areas of ocean science. Overlapping topics may indicate fruitful areas for establishing or supporting networks of researchers with related and complementary interests. For example, several themes recognize the lack of information in relatively inaccessible regions including the Arctic Ocean, Southern Ocean, and deep ocean (waters deeper than 200 m and the seafloor), yet the need to understand these regions becomes ever more critical as climate and human stressors drive changes in physical, chemical, and biological processes. Similarly, several themes touch on the demand for information to expand, support, and manage a sustainable blue (or ocean) economy. This includes increasing access to data and models as well as their analysis and interpretation. Application of the information to support sustainable development will require workforce training as well as increased ocean literacy, not just of ocean processes and biology but also of the importance of the ocean for the economy and the cultural heritage of coastal communities.

Identification of themes is only one step toward realizing the goals of science for the "ocean we want" during the UN Ocean Decade. Progress will depend on the development of innovative programs that attract substantial research funding. The committee suggests that next steps involve a series of workshops—first to focus on components of the themes and then to meld these components together into a compelling, overarching program. These workshops will include expertise from outside of the immediate ocean science community to cross-fertilize ideas from other complementary disciplines drawn from the breadth of natural and social science and engineering experts available through the various activities (boards, roundtables, and standing committees) of the National Academies. In addition, representatives from stakeholder groups, including industry and funding organizations (public and private), will be encouraged to attend and participate in the workshops to better represent their interests in program development. For discussion of each topical theme, representatives from the two foundational themes will be present to ensure that the best practices and approaches developed for these themes are effectively integrated.

The UN Ocean Decade represents a global opportunity to advance ocean science, implement sustainable practices, and raise awareness of humanity's dependence on the ocean. It is also an opportunity for the United States to demonstrate international leadership in furthering efforts to ensure that the ocean continues to support the well-being of communities at home and worldwide, for both current and future generations. The two foundational and four topical themes recommended by the committee, if further developed and funded, have the potential to greatly advance our understanding of the ocean—its processes, resources, and values—to enable "the ocean we need for the future we want." The UN Ocean Decade also provides an opportunity to rethink how science is conducted with an emphasis on open engagement of a broad, inclusive community in the development and conduct of research matched with greater access to data, data products, and models. Building on the efforts of the U.S. ocean community who submitted Ocean-Shots, activities endorsed by the UN Ocean Decade process, and many complementary efforts such as OceanObs'19 and the High Level Panel reports, the committee offers these themes as the next step toward realizing the promise of the UN Ocean Decade.

1

Introduction

In December 2017, the United Nations (UN) General Assembly proclaimed 2021–2030 the "Ocean Decade," officially the UN Decade of Ocean Science for Sustainable Development, providing international recognition at the highest levels of government of the importance of the ocean and ocean science for achieving the United Nations' Sustainable Development Goals.[1] The vision for the UN Ocean Decade is "the science we need for the ocean we want." The concept of the UN Ocean Decade is to harness the collective power of the global community to address increasing challenges for sustainable development in areas such as food security, renewable energy, pollution, and climate change. The mission for the UN Ocean Decade, "to catalyse transformative ocean science solutions for sustainable development, connecting people and our ocean," emphasizes the innovative nature and collaborative potential of this undertaking (UNESCO-IOC, 2021a).

In advance of the launch of the UN Ocean Decade in January 2021, the United Nations Educational, Scientific and Cultural Organization's Intergovernmental Oceanographic Commission (UNESCO-IOC) released the UN Ocean Decade Implementation Plan that outlines a framework for coordinating and promoting the UN Ocean Decade in areas around the globe, and empowering individuals and groups to engage, plan, and implement UN Ocean Decade goals in a shared way (UNESCO-IOC, 2021a). Included in the Implementation Plan are the UN Ocean Decade "Challenges," "Outcomes," and "Actions" (see Box 1.1) and recommendations for the development of National Decade Committees, which "will be essential elements to engage national stakeholders and facilitate

[1] See https://sdgs.un.org/goals.

BOX 1.1
The UN Ocean Decade Definitions[a]

Outcomes: the endpoints to achieving "the ocean we want."

Challenges: "form the highest level of this framework and represent the most immediate and pressing priorities for the Ocean Decade. They aim to unite Decade partners in collective action at the global, regional, national and local scales and will contribute to the achievement of the Ocean Decade outcomes, thus shaping the overall contribution of the Decade to the 2030 Agenda and other policy frameworks."

Actions: "the tangible initiatives that will be carried out across the globe over the next ten years to fulfill the Ocean Decade vision. They will be carried out by a wide range of proponents, including research institutes and universities, governments, UN entities, intergovernmental organizations, other international and regional organizations, business and industry, philanthropic and corporate foundations, NGOs [nongovernmental organizations], educators, community groups or individuals."[b]

[a] UNESCO-IOC, 2021a.
[b] Tangible initiatives are categorized as programs, projects, activities, and/or contributions.

national contributions to the Decade as well as to promote awareness and interest" and "be key in linking national action to the international UN Ocean Decade framework" (UNESCO-IOC, 2021b).

Since the publication of the Implementation Plan, UNESCO-IOC has developed and launched an interactive platform—the UN Ocean Decade Laboratories—the goals for which are to catalyze action for the UN Ocean Decade, showcase UN Ocean Decade actions, strengthen dialogue, and enhance communication and outreach.[2] Each laboratory focuses on one of the seven outcomes of the UN Ocean Decade. To date, the following laboratories have taken place: An Inspiring and Engaging Ocean (July 2021), A Predicted Ocean (September 2021), A Clean Ocean (November 2021), A Healthy and Resilient Ocean (March 2022), A Safe Ocean (April 2022). The remaining two outcomes (A Productive Ocean and An Accessible Ocean) will be the topics of UN Ocean Decade Laboratories planned for the remainder of 2022.

At the request of the National Science and Technology Council's Subcommittee on Ocean Science and Technology (SOST),[3] the National Academies of Sciences, Engineering, and Medicine (the National Academies) established the

[2] See https://www.oceandecade-conference.com/en/ocean-decade-laboratories.html.
[3] See https://www.noaa.gov/ocean-science-and-technology-subcommittee.

U.S. National Committee for the UN Decade of Ocean Science for Sustainable Development (U.S. National Committee) to inspire U.S. contributions to the UN Ocean Decade and to serve as the informational hub for U.S. Ocean Decade–related activities. The communications role of the U.S. National Committee has been achieved through its website,[4] a network of more than 80 Nexus organizations, newsletters, an events calendar, social media, and a series of webinars including a kick-off meeting in February 2020 that had more than 1,000 registrants. The U.S. National Committee also facilitates the development of partnerships in support of Ocean Decade–related ocean science activities. It has actively sought the engagement of early career ocean scientists through the recruitment and selection of liaisons to the U.S. National Committee[5] and the involvement of youth ages 14–25 through the U.S. Youth Advisory Council for the UN Ocean Decade.[6]

To engage and inspire the U.S. ocean scientific community, the U.S. National Committee issued a call for "Ocean-Shots," defined as ambitious, transformational research concepts that draw inspiration and expertise from multiple disciplines and fundamentally advance ocean science for sustainable development. The U.S. National Committee held a public virtual meeting in the fall of 2020 to announce the call for "Ocean-Shots," and publicized it on the U.S. National Committee website, in the U.S. National Committee newsletter, and through announcements shared by federal agencies and nonfederal organizations. With the encouragement of the SOST, the U.S. National Committee's vision was to take advantage of the visibility and momentum of the UN Ocean Decade to go beyond "business as usual" and foster the creation of programs of such broad-reaching scale and relevance that they would capture the imagination of the science community, funders, the public, and policy makers, much like the "moonshot" mission to the Moon of the 1960s.[7]

The call for Ocean-Shots offered a series of criteria for the submissions. These included demonstrating potential transformative impact; engaging scientific, technical, and other sectors outside the traditional ocean sciences; building capacity and a cadre of next-generation ocean scientists; and directly addressing at least 1 of the 10 "Challenges" set forth in the UN Ocean Decade's Implementation Plan. The U.S. National Committee anticipated that some of the ideas submitted as Ocean-Shots could be developed into projects or programs for submission to the UN Ocean Decade for endorsement. Consequently, the criteria also included identification of opportunities for international participation and collaboration. The Ocean-Shots were not meant to duplicate the UN Ocean Decade call for actions; rather, the Ocean-Shots were meant to provide the impetus

---

[4] See www.nationalacademies.org/oceandecadeus.

[5] See https://www.nationalacademies.org/our-work/us-national-committee-on-ocean-science-for-sustainable-development-2021-2030#sl-three-columns-c494ad96-a895-4f0f-962d-11563f650d8d.

[6] See https://h2oo.org/us-yac-for-un-ocean-decade.

[7] See https://www.jfklibrary.org/visit-museum/exhibits/past-exhibits/moon-shot-jfk-and-space-exploration.

for the development of new projects and programs over the course of the decade or, in some cases, provide synergistic activities that involved and complemented UN Ocean Decade–endorsed actions that include significant U.S. ocean community input. The response to the call for Ocean-Shots yielded more than 100 submissions, demonstrating the breadth, skill, creativity, and enthusiasm of the U.S. marine science and engineering communities and their willingness to build on the themes of the UN Ocean Decade and engage in transformative science in support of sustainable development.[8]

Exciting concepts were submitted in response to the call for Ocean-Shots (now accepted on a rolling basis). However, neither the UN Ocean Decade nor the U.S. National Committee are sources of funding that could bring the concepts to fruition. Recognizing the immense potential of the Ocean-Shots to stimulate U.S. ocean science activities in support of the UN Ocean Decade, the U.S. National Committee discussed possible next steps to further development of these ideas with the co-chairs of the SOST, represented by the White House Office of Science and Technology Policy, the Office of Naval Research, the National Oceanic and Atmospheric Administration, and the National Science Foundation. This conversation led to a request from the National Aeronautics and Space Administration, on behalf of the SOST, for a consensus study to identify between three and five themes, as described in the Statement of Task (see Box 1.2). The Committee on Cross-Cutting Themes for U.S. Contributions to the Ocean Decade was appointed to undertake this task in October 2021 by the National Academies based on the membership of the U.S. National Committee and including four early career liaisons (see Appendix B). To ensure that the ocean community had an opportunity to contribute to and comment on potential themes, the committee posted its draft list of themes on the Ocean Decade U.S. website for a 2-week public comment period. These comments were reviewed by the committee and contributed to the final formulation of the themes as presented in this report.

The Statement of Task explicitly asks the committee to build its recommendations around the Ocean-Shot submissions. At the same time, recognizing that this effort is in support of the UN Ocean Decade, the recommended themes should support the UN Ocean Decade priorities as described in the Implementation Plan (UNESCO-IOC, 2021a; see Table 1.1). As a U.S. effort, the SOST also asked that the recommended themes complement U.S. ocean science priorities as described in *Science and Technology for America's Oceans: A Decadal Vision* (SOST, 2018; see Table 1.1). A comparison of the seven UN Ocean Decade objectives with the five identified U.S. ocean priorities (see Table 1.1) clearly shows that the visions of these two efforts move us closer to "the ocean we want." The two sets of priorities complement each other, with the UN Ocean Decade priorities focusing on the health of the ocean and the U.S. priorities focusing on the health and well-being

[8] See https://www.nationalacademies.org/our-work/us-national-committee-on-ocean-science-for-sustainable-development-2021-2030/ocean-shot-directory.

BOX 1.2
Statement of Task

Based on the body of submissions to the call for Ocean-Shots as part of the National Academies project on U.S. Contributions to the Ocean Decade, the ad hoc consensus committee will identify 3–5 cross-cutting themes that incorporate the most promising and innovative research concepts. Specifically, the committee will examine how each theme aligns with the overall goal of the UN Ocean Decade in supporting ocean science for sustainable development with potential for generating future UN Decade Programmes. In addition, the committee will connect the themes to U.S. ocean priorities, as identified in documents such as the National Science and Technology Council, Subcommittee on Ocean Science and Technology document *Science and Technology for America's Oceans: A Decadal Vision* (2018). The themes identified by the committee will address compelling areas for public and private sector investment and provide opportunities for inter- and multi-disciplinary activities in support of ocean science. The committee will prepare a short report that identifies the 3–5 themes and describes each briefly according to the criteria outlined above.

of the communities that depend on a healthy ocean. Finally, the Statement of Task specifies that the themes identified by the committee address compelling areas for public- and private-sector investment and provide opportunities for inter- and multidisciplinary activities in support of ocean science. The committee interprets this statement as an indication that, with compelling themes for the UN Ocean Decade, the agencies represented in the SOST will be encouraged to seek the means to support research addressing these themes. The identified themes also align with other efforts to advance sustainable ocean science, including the goals described in the OceanObs'19 Conference Statement (OceanObs'19, 2019) and the transformations emphasized in The High Level Panel for a Sustainable Ocean Economy's (i.e., the Ocean Panel[9]) new ocean action agenda, *Transformations for a Sustainable Ocean Economy: A Vision for Protection, Production and Prosperity* (Ocean Panel, 2020). The themes complement these ongoing efforts and foster a multidisciplinary approach for the initiation of research activities to advance the Sustainable Development Goals for the UN Ocean Decade.

Each theme begins with an overview of the issues to be addressed, providing context and explaining how work under the theme would not only contribute to but also significantly advance ocean science for sustainable development. The themes were developed within the context of the UN Ocean Decade initiatives, as described in the Implementation Plan. Consequently, the committee recognized

[9] See https://www.oceanpanel.org/about; https://www.oceanpanel.org/ocean-policy.

TABLE 1.1 UN Ocean Decade Challenges, UN Ocean Decade Outcomes, and U.S. Ocean Priorities

UN Ocean Decade Challenges (i.e., Ten challenges for collective impact in the Decade ahead)	UN Ocean Decade Outcomes (i.e., Seven outcomes to describe the "ocean we want" at the end of the UN Ocean Decade)	U.S. Ocean Priorities (i.e., Five goals to advance U.S. Ocean Science and Technology and the Nation in the coming decade)
1. Understand and beat marine pollution	1. A clean ocean where sources of pollution are identified and reduced or removed	1. Understand the Ocean in the Earth System
2. Protect and restore ecosystems and biodiversity	2. A healthy and resilient ocean where marine ecosystems are understood, protected, restored and managed	2. Promote Economic Prosperity
3. Sustainably feed the global population	3. A productive ocean supporting sustainable food supply and a sustainable ocean economy	3. Ensure Maritime Security
4. Develop a sustainable and equitable ocean economy	4. A predicted ocean where society understands and can respond to changing ocean conditions	4. Safeguard Human Health
5. Unlock ocean-based solutions to climate change	5. A safe ocean where life and livelihoods are protected from ocean-related hazards	5. Develop Resilient Coastal Communities
6. Increase community resilience to ocean hazards	6. An accessible ocean with open and equitable access to data, information and technology and innovation	
7. Expand the Global Ocean Observing System	7. An inspiring and engaging ocean where society understands and values the ocean in relation to human well-being and sustainable development	
8. Create a digital representation of the ocean		
9. Skills, knowledge and technology for all		
10. Change humanity's relationship with the ocean		

SOURCES: SOST, 2018; UNESCO-IOC, 2021a.

the importance of including a section specifically to identify the UN Ocean Decade outcomes and challenges that would be addressed by the pursuit of these themes.

Critically, the themes were inspired by the body of Ocean-Shots submitted by the U.S. ocean science community. Through review of the more than 100 Ocean-Shots, the committee identified common threads linking many of the submissions. Starting with these "threads," the committee narrowed the list to the required three to five themes specified in the Statement of Task. Two of the emergent themes were viewed as fundamental to supporting the execution of more topical research efforts. These are identified as the foundational themes—An Inclusive and Equitable Ocean and An Ocean of Data. In addition, the committee identified four topical themes: The Ocean Revealed, The Restored and Sustainable Ocean, Ocean Solutions for Climate Resilience, and Healthy Urban Seas. Ocean-Shots that contribute to these themes are listed in the next chapter. Although these Ocean-Shots have high relevance for a specific theme, the list is not meant to be exclusive or exhaustive; some Ocean-Shots fit multiple themes. In addition, Chapter 2 lists UN-endorsed programs and projects relevant to the theme because many U.S. scientists have already engaged in the UN Ocean Decade through submissions at the international level.

Following the listing of relevant Ocean-Shots and UN-endorsed programs, each theme has a section titled "Potential Research Elements." This section, derived mostly from concepts presented in relevant Ocean-Shots, offers potential research projects that may become part of the overall theme should it be funded. It is meant to demonstrate that while the theme is being presented as a framework, many key building blocks have already been crafted. It is critical, however, to recognize that these research elements are not meant to represent the full scope of the program but rather to serve as examples drawn from disparate Ocean-Shots with relevance to the theme. Should it be decided that further investment is warranted, the critical role of defining the appropriate research building blocks for the theme will occur through a series of focused workshops and other discussions specifically designed around theme components, as described in the overview of "Potential Next Steps" below.

Under the "Potential Next Steps" sections of Chapter 2, the committee outlines a series of workshops that focus on components of the theme and then bring the various interested communities (e.g., researchers, managers, engineers, funders, stakeholders) together to structure an overarching program. The workshops will provide an opportunity for developing interdisciplinary teams that represent various science, technology, engineering, mathematics, and social sciences disciplines, including early career and historically marginalized groups, and come from a variety of backgrounds (e.g., government, academia, industry, nonprofits, foundations) to initiate the development of plans with enough specificity to merit decisions on the viability of and potential for funding. Workshops, pending funding from the public or private sector, would initially draw from the cohort of investigators who submitted Ocean-Shots relevant to that theme, with the investigators having the option to attend the workshop most relevant to their interests. However, workshop participation would be broadened by an active effort to ensure the inclusion of early career scientists, underrepresented groups, Indigenous representatives, stakeholders, the private sector, and research funders.

With input from the committee, gaps in the existing Ocean-Shots that prevent achievement of the overarching goals of the theme will be identified and a diverse group of experts will be solicited to fill those gaps. The committee will call on other National Academies boards to identify experts from other disciplines to enable an interdisciplinary approach for addressing the themes. In each case, representatives from the two foundational themes will be present to ensure that the best practices and approaches developed by these themes are integrated into the theme's development. The goal for these workshops is to produce well-developed research plans with enough detail to allow the SOST and others to evaluate the feasibility of making substantial investments into these decadal-scale programs that are designed to address key research and actions necessary to ensure a safe, healthy, and sustainable ocean.

Finally, a short section on "Defining Success" is presented for each theme in Chapter 2. This section captures the initial thoughts of the committee on what the state of the ocean and ocean science might be if the objectives outlined in each theme are met. In some cases, objectives are measurable; in other cases, they are less concrete. Through the process of the "Next Steps" workshops, a much clearer picture of the metrics for success for each theme will be defined. In all cases the definitions of success are aspirational—some may not be achieved by 2030, but much will be gained trying. Additionally, the UN Ocean Decade is a construct; the problems of ocean sustainability are real and will need to be addressed far beyond the end of the decade. By increasing public awareness of the critical functions of the ocean, the UN Ocean Decade will increase the long-term interest in sustainable ocean practices.

In the final chapter, the committee provides concluding thoughts on how the described themes could help advance the U.S. contributions to the UN Ocean Decade and next steps to develop the themes into mature research programs.

2

Themes for U.S. Contributions to the Ocean Decade

The United Nations (UN) Decade of Ocean Science for Sustainable Development 2021–2030 (UN Ocean Decade) offers what is perhaps a once-in-a-lifetime opportunity to engage the global community in recognizing the critical importance of the ocean and ocean science to the long-term well-being of humankind. The U.S. National Committee for the UN Decade of Ocean Science for Sustainable Development (U.S. National Committee) was established as an informational focal point for UN Ocean Decade activities occurring across the United States and, more importantly, to inspire the U.S. community to propose bold, creative, and audacious new ways to advance ocean science in support of sustainable development. The call for "Ocean-Shots" was designed to initiate this engagement with the ocean science community.

The federal agencies represented on the Subcommittee on Ocean Science and Technology subsequently asked the U.S. National Committee to identify cross-cutting themes that encompass the most promising and innovative research concepts of the submitted Ocean-Shots, and complement the goals of the UN Ocean Decade, including the UN-endorsed actions, and established U.S. ocean priorities. The results of the committee's deliberations are presented below as two foundational themes and four topical themes. These themes were previously posted for public comment and presented in a virtual public meeting on November 12, 2021. The committee received and reviewed the feedback on the draft themes, and these comments informed the final versions of the themes presented here. These themes are not meant to encompass every important topic of ocean research; instead, they represent a subset of promising areas built around submitted Ocean-Shots and are meant to inspire additional investment into research that will support the overarching sustainability goals of the UN Ocean Decade for years to come.

The foundational themes (An Inclusive and Equitable Ocean and An Ocean of Data) focus on the underlying infrastructure of science, ensuring that information resources are accessible, applicable, and appropriate for diverse communities while recruiting the range of talents needed to solve the ocean's most pressing problems. As these themes are developed, protocols and best practices will be established for application in each of the four topical themes. The topical themes (The Ocean Revealed, The Restored and Sustainable Ocean, Ocean Solutions for Climate Resilience, and Healthy Urban Seas) address fundamental aspects of sustainability and exploration and offer frameworks under which exciting new research programs can be further refined and developed. Note that these themes are offered as frameworks; the core of what is needed to develop these into tangible and fundable programs will be established through focused subsequent activities with researchers and users.

The committee proposes to convene workshops first on the foundational themes to establish best practices. Each follow-up activity associated with a topical theme will start with planning to incorporate best practices identified during the workshops for the two foundational themes.

FOUNDATIONAL THEMES

An Inclusive and Equitable Ocean

Overview of Theme

The UN Ocean Decade launches at a unique moment when society brings renewed attention and dedication to creating an environment that is truly inclusive, where work is framed with a larger sense of purpose, participation is invited with an acknowledgment that there are gaps in one's understanding and experience, and consideration is given to all ideas and viewpoints. It represents an opportunity for positive change in broadening who participates in the ocean enterprise and how ocean research is conducted across all aspects of ocean studies such as the natural and social sciences; ocean literacy, education, and professional development; ocean governance, policy, and management; ocean industry; and others. The best and the brightest from a variety of backgrounds will be needed to solve critical questions in ocean sciences.

An elemental aspect of the UN Ocean Decade is its recognition that increasing awareness, understanding, and sharing of all the ocean has to offer humanity can only be achieved through the integrated involvement of a diverse and representative ocean community. Inclusivity means communities can share their knowledge and their local understanding of the natural environment, thus contributing to the science instead of simply learning about it. Talented students from diverse backgrounds should be recruited into the field in order to solve critical questions in ocean science. Diversity enhances creativity, and innovation is driven by diverse teams that facilitate novel thinking and adaptability (Hofstra et al., 2020). Equity, inclusiveness, respect, fairness, and scientific integrity are core principles of the UN Ocean Decade and as such permeate all activities; in the words of the UN Ocean Decade, "no one is left behind" and all are empowered to contribute to the ocean science enterprise.

Initial efforts on this foundational theme will define an inclusive and equitable ocean and establish a framework and methodology to incorporate core principles into each topical theme. The primary outcome of this theme will be to elevate equity and inclusion as a priority for the UN Ocean Decade, and help define and guide implementation of inclusive and equitable ocean science practices throughout the topical themes.

Under this theme, additional topics for research may be identified that have applications to ocean sustainability, such as studies on the variety of barriers encountered by individuals from underrepresented groups in pursuing a career in ocean studies. This could be followed by an effort to identify effective practices that empower students to seek and succeed in ocean careers. Other activities could include the development of programs for engaging underserved coastal communities to identify needs and contribute to gathering information on topics such as securing benefits from growth in the blue economy, coastal restoration, or climate adaptation.

An Inclusive and Equitable Ocean will further the development of approaches that span the scientific, technical, policy, management, Indigenous, and stakeholder communities to ensure involvement of the full diversity of society in setting science priorities for sustainable ocean development and ocean knowledge generation.

Decade Challenges and Outcomes Addressed

This theme directly addresses UN Ocean Decade Outcome 6: "An accessible ocean with open and equitable access to data, information and technology and innovation" and Outcome 7: "An inspiring and engaging ocean where society understands and values the ocean in relation to human well-being and sustainable development" (UNESCO-IOC, 2021a). It also directly addresses UN Ocean Decade Challenge 9: "Skills, knowledge and technology for all - Ensure comprehensive capacity development and equitable access to data, information, knowledge and technology across all aspects of ocean science and for all stakeholders" and Challenge 10: "Change humanity's relationship with the ocean - Ensure that the multiple values and services of the ocean for human well-being, culture, and sustainable development are widely understood, and identify and overcome barriers to behaviour change required for a step change in humanity's relationship with the ocean."[1] In developing principles and guidelines for addressing these chal-

[1] See https://www.oceandecade.org/challenges.

lenges and outcomes, this theme will help inform the conduct of other UN Ocean Decade research programs and guide the application of knowledge to action.

Connections to Ocean-Shots, UN Ocean Decade Actions, and U.S. Ocean Priorities

This foundational theme builds on expert discussions organized by the National Academies' Ocean Studies Board to examine diversity, equity, inclusion, justice, and belonging in ocean studies. It also addresses nearly 20 Ocean-Shot submissions and complements UN Ocean Decade–endorsed actions on the critical issue of involving diverse communities in ocean-related research; incorporating broad cultural values; and embracing the principles of justice, equity, diversity, and inclusiveness (see Table 2.1). The value of engagement of diverse communities is recognized as an area of immediate opportunity as described under "Improve data integration in decision-support tools" in *Science and Technology for America's Oceans: A Decadal Vision* (U.S. Decadal Vision; SOST, 2018). From the development of ocean science knowledge networks to the creation of tools for ocean science education, outreach, and engagement, the community clearly recognizes the need for change and submitted a range of ideas in the Ocean-Shots to achieve that change.

Potential Research Elements

- Identify methodology related to
  - broadening approaches to knowledge generation and use, including co-development and equitable exchange, to incorporate different cultures, disciplines, and values;
  - developing strategies to relate place-based and qualitative knowledge to the more quantitative and large-scale, big data approaches and ways to incorporate different types of data, considering different types of evidence that are important for the co-creation of knowledge;
  - increasing scientific and technical capacity from local to regional scales, and across sectors to include broader participation;
  - improving access to data, information, technology, and research infrastructure in underserved communities and countries; and
  - building appreciation of the ocean's economic, social, and cultural values to public and individual health and well-being, and sustainable development through promotion of ocean literacy.
- Identify equity and inclusion-specific research areas that can inform ocean research, such as a place-based analysis of past community-engagement programs, comparisons of ocean and land-based governance approaches, and studies on the effectiveness of citizen/community science initiatives in building confidence in scientific information.
- Develop principles, guidance, and an action plan for incorporating these

TABLE 2.1 Connections of An Inclusive and Equitable Ocean to Ocean-Shots and UN Ocean Decade Actions

Title
Ocean-Shots
FantaSEAS Project: Incorporating Inspiring Ocean Science in the Popular Media
EquiSea: The Ocean Science Fund for All
The Ocean Decade Show
The Estuarine Ecological Knowledge Network (EEKN): The View from SE Louisiana and Future Prospects
TRITON: A Social Media Network for the Ocean
Ocean Technology Field Academy
Small Islands, Big Impact
ICOFS (Integrated Coastal Ocean Forecast Systems)
An Ocean Science Education Network for the Decade
Building Ocean Collaborations
Envisioning an Interconnected Ocean: Understanding the Links Between Geological Ocean Structure and Coastal Communities in the Pacific
Ocean Memory Project: A Cross-Disciplinary Approach to Global Scale Challenges
The 4Site Pacific Transect Collaborative
Just, Equitable, Diverse, and Inclusive (JEDI) Aquanautics: Democratizing Innovation in the Networked Blue Economy
Revolutionizing Coastal Ocean Research through a Novel Share Model for the Long-term Sustainability of Humanity
An Ocean Corps for Ocean Science (also a UN Ocean Decade Endorsed Programme)
UN Ocean Decade Endorsed Actions[a]
Ocean Voices: Building transformative pathways to achieve the Decade's outcomes (Decade Programme 16)
An Ocean Corps for Ocean Science (Decade Programme 9)
AGU's Mentoring365: UN Decade of Ocean Sciences (Decade Contribution 226)
A Multi-Dimensional and Inclusive Approach for Transformative Capacity Development (CAP-DEV 4 the Ocean) (Decade Project 39)

NOTES: See Appendix A, Table A.1, for a description of each. AGU, American Geophysical Union.
[a] See https://www.oceandecade.org/decade-actions.

elements into each topical theme and using them to inform future activities in support of the UN Ocean Decade.

- Identify and compile best practices for equitably applying and adapting knowledge, resources, and activities to meet the needs of diverse communities across scales.

Potential Next Steps

If funding is received, a workshop with diversity, equity, inclusion, and environmental justice experts; social and natural scientists; and other interested scholars and practitioners will be organized as the initial activity to take place after the publication of this report. This workshop will help to identify approaches for broader involvement of diverse communities in contributing to and determining needs for ocean science for sustainable development and to guide the incorporation of effective practices into each of the topical themes. Resources and guidelines for furthering inclusion and equity will be presented. Examples of programs that have succeeded in or aspired to address various aspects of this issue will be discussed. The goal will be to identify key actions for a more inclusive and equitable ocean and how to implement them in the topical themes and future activities in support of the UN Ocean Decade. Some of the actions may include the following:

- Early engagement and co-development of research topics and sustained involvement of diverse stakeholders in the study process and outcomes;
- Community engagement in data collection, analysis, interpretation, and development of sustainable solutions;
- Effective approaches for presenting and communicating scientific information that cross natural science, social science, and humanities disciplines;
- A knowledge repository, including training materials and best practices, to inform and support equitable and inclusive practices in the ocean community; and
- Creation of a Leadership Learning Circle for learning and sharing among organizations and people who are leading the vanguard in diverse, equitable, and inclusive ocean practices and those seeking guidance in launching similar efforts.

The proposed workshop will refine this list, set a foundation, and chart the course for its implementation into each topical theme.

Defining Success

The committee's vision is that the full, diverse spectrum of society recognizes the importance of studying the ocean and working together to achieve sustainability. Scientific decisions will be made with interdisciplinary contributions, representing the participation of many groups and recognizing the value of local knowledge and different knowledge systems. All members of the community come together—each with a voice, a sense of belonging, and a shared purpose—to use ocean science to make a positive impact on people's lives.

The UN Ocean Decade will be distinguished by the level of engagement of communities outside the realm of traditional academic research, including the following:

- All four topical themes actively involve the broad community in the development of their research plans, as well as create, implement, and adaptively monitor the outcomes of an equity and inclusion action plan.
- All communities feel welcome to engage in Ocean Decade activities, with recognition of their interests, concerns, and active participation.
- Metrics are developed, applied, and documented such that each community can track meaningful points of progress in creating an equitable and inclusive Decade of Ocean Science for Sustainable Development.
- Ocean science, knowledge, educational materials, and social engagement are made accessible and broadly available through an open access platform.

An Ocean of Data

Overview of Theme

An explosion in ocean data is taking place from new ocean observing systems, an ocean "Internet of Things" (Lueth, 2014; Waterston, n.d.), numerical modeling output, citizen/community science, information from smart technologies (e.g., smartphones), and social media postings. These data sources include those beneath the surface of the ocean as well as those on the surface, in the air, and in space. The diversity of data types is growing as fast as the volume of data. The parallel development of increased computational capacity, artificial intelligence (AI), machine learning (ML), and other emerging processing techniques creates opportunities for assimilating and querying multiple data streams in new ways and at unprecedented speeds. This offers the prospect of a better understanding of the ocean and greater predictive capacity to inform safety at sea and to evaluate alternatives for managing ocean resources sustainably. Moreover, new sensor networks that rely on robotics and low-cost sensors are emerging that are allowing broader coverage of ocean observations and access by a wider range of developers and users.

There have been many data initiatives over the past several decades focusing on actions such as shared databases, interoperability, data dictionaries, and common formats. However, none of them truly achieved their ambitious aims, and the result is that we now have numerous web portals without any consistent architectural design. Although these web portals focus on the needs of the data providers, they lack consideration for how the data will actually be used. An alternate approach for the UN Ocean Decade is to follow the vision of Buck et al. (2019) who proposed a democratized structure where users create their own data and knowledge services using tagged and labeled data. Moving toward an open science community of practice (NASEM, 2018b) and harnessing modern information technology (Gentemann et al., 2021) will create frameworks and workflows that have the potential to overcome emerging big data challenges (Ramachandran et al., 2021).

Historically, ocean data have been held in national, regional, institutional, or even individual archives. With few exceptions, ocean data have been prohibitively expensive to collect, often requiring ship-time and hence limiting coverage in both time and space. Over the past several decades there has been a major shift toward the principle of open access for data collected with government funds, but because of the enormous range of data types, quality, and formats, and the enormous range of institutions and individuals who hold data, accessibility and usability are far from optimal. While some types of data, especially physical measurements, can be cataloged using relatively straightforward criteria, observations on some biological properties and human dimensions often have greater variability and may require additional contextual data. The database architecture could unintentionally exclude these types of data, thereby narrowing the range of data archived. As a result, even today, a significant fraction of existing ocean data remains locked away, unorganized, and/or closely held by government agencies, companies, resource users, or researchers. In addition, many data sets were collected and stored in ways that were specifically designed for, and suited to, the original users' needs or were constrained by technical or operational challenges at the time. This has sometimes created distrust within Least Developed Countries (LDCs) and Small Island Developing States (SIDS) owing to the perception that wealthy nations and organizations collect and use data for their own reputational or financial gain without recognizing the needs and contributions of LDCs and SIDS in a meaningful way. These factors have led to the lack of more "universal" and broadly applicable data sets, which presents a substantial roadblock for addressing the challenges and reaching the objectives of the UN Ocean Decade.

The many examples of siloed data make it difficult to combine insights and data across programs that are needed for a more interdisciplinary and holistic understanding of the interaction between societal interests and ocean conditions and trends. Such a democratization of data will be required to unleash the full potential of AI and ML. The unprecedented volumes of new data being acquired with increasing speed make it prohibitively expensive to move data from their sources (e.g., including the output of large-scale global ocean and climate models) to local users who wish to perform analyses or create services. For example, the effective application of the "digital twin" approach (a model designed to represent a component of a physical system, such as an ocean basin, that is "tuned" with data from the physical twin) will depend on greater access to both data and computational capacity. This situation illustrates the need for cloud-based and remote computational engines to access and analyze these remote data.

Because future data needs and knowledge services cannot be anticipated, democratized data access and services that are driven by a user-centric and decentralized process will be required before a vision of highly adaptive and dynamic ocean management approaches can be achieved. Scientists need to rethink the intersections among observing systems, information systems, and knowledge systems that are focused on delivering services and solutions, not just data. New approaches, such as placing tagged or labeled data within data "lakes," focus on implementing easy-to-use and extensible[2] interfaces for adding new data types as well as for creating new services based on these data.

The UN Ocean Decade is a unique and timely opportunity to foster the coordination needed to make ocean data and computational services available to meet the pressing needs of the future. The UN Ocean Decade could enable findable, accessible, interoperable, and reusable (FAIR; Wilkinson et al., 2016) and collective benefit, authority to control, responsibility, and ethics (CARE; GIDA, 2019) data systems, including standardization and calibration to maximize the value of the data to a broad community of users. This includes standards or best practices for assessment and reporting of uncertainty regarding the reliability of the data to guide users of the information. It will be necessary to change norms and expectations to liberate data and to foster the innovations that will put actionable information in the hands of managers, users, and stakeholders. In partnership with existing UN Ocean Decade actions and private-sector initiatives, the United States could play a key role in the creation of an open, actionable, and equitable digital ecosystem for ocean knowledge across all ocean arenas.

As a foundational theme, An Ocean of Data will improve data availability and access through the development of a framework for implementing FAIR and CARE data principles and the creation of a path toward a digital ecosystem that delivers ubiquitous compute-intensive data services. Consideration of these digital ocean outcomes will also assist in the operationalization of each topical theme.

Decade Challenges and Outcomes Addressed

While this theme supports all seven of the UN Ocean Decade outcomes, it most directly fulfills Outcome 6: "An accessible ocean with open and equitable

[2] The ability to extend a system to add new capabilities or functions, generally in reference to software.

access to data, information and technology and innovation" (UNESCO-IOC, 2021a). It also directly addresses UN Ocean Decade Challenge 7: "Expand the Global Ocean Observing System - Ensure a sustainable ocean observing system across all ocean basins that delivers accessible, timely, and actionable data and information to all users" and UN Ocean Decade Challenge 9: "Skills, knowledge and technology for all - Ensure comprehensive capacity development and equitable access to data, information, knowledge and technology across all aspects of ocean science and for all stakeholders."[3] Development of data and computational systems will support the other challenges that require improving understanding, generating new knowledge, and enhancing predictive capacity. Additionally, there is a need to evaluate and quantify the quality and usability of various data and models.

Connections to Ocean-Shots, UN Ocean Decade Actions, and U.S. Ocean Priorities

Nearly 10 Ocean-Shots (see Figure 2.1), the UN Ocean Decade Implementation Plan, and UN Ocean Decade–endorsed actions specifically addressed and helped guide the committee's development of this foundational theme (see Table 2.2). The submissions received ranged from digital integration of ocean observing systems to ocean exploration using AI. This theme also directly addresses three of the five immediate research needs identified by the National Science and Technology Council's (NSTC's) U.S. Decadal Vision for the oceans (SOST, 2018):

1. Fully integrating data-intensive approaches in Earth system science,
2. Advancing monitoring and predictive modeling capabilities, and
3. Improving data integration and interoperability for decision-making.

Potential Research Elements

- Develop a new vision for ocean data that enables users to design and deploy knowledge services that meet their needs with greater flexibility than existing options by
    - identifying principles and a framework that will enable adaptive and flexible approaches to the management of our ocean, with a particular focus on democratization of and open access to user-created knowledge services;
    - identifying best practices for evaluating quality and fostering quantification of data and model uncertainty; and

[3] See https://www.oceandecade.org/challenges.

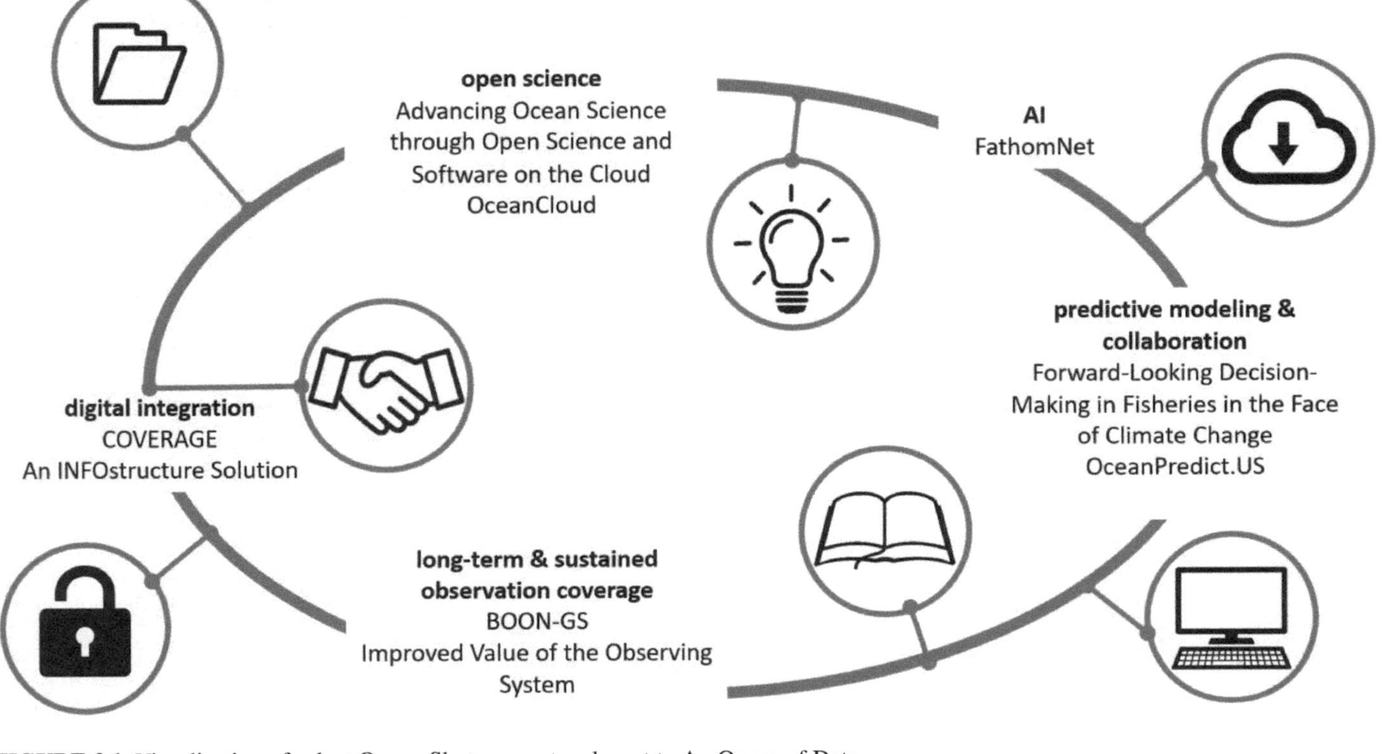

FIGURE 2.1 Visualization of select Ocean-Shot concepts relevant to An Ocean of Data.
NOTE: Acronyms are defined in Table 2.2.

TABLE 2.2 Connections of An Ocean of Data to Ocean-Shots and UN Ocean Decade Actions

Title
Ocean-Shots
Advancing Ocean Science through Open Science and Software on the Cloud
Boundary Ocean Observation Network for the Global South (BOON-GS)
COVERAGE: Next Generation Data Service Infrastructure for a Digitally Integrated Ocean Observing System in Support of Marine Science and Ecosystem Management
FathomNet: Exploring Our Ocean Using Artificial Intelligence
Forward-Looking Decision Making in Fisheries in the Face of Climate Change
Improved Value of the Observing System through Integrated Satellite and in situ Design
OceanCloud: Transforming Oceanography with a New Approach to Data and Computing
OceanPredict.US
An INFOstructure solution to the socio-ecological hazards of coastal flood control infrastructure
UN Ocean Decade Endorsed Actions[a]
Deep Ocean Observing Strategy (Decade Programme 129)
The World Ocean Database Programme (WODP) (Decade Contribution 122)
WOC SMART Ocean-SMART Industries (SO-SI): Science/Industry Partnerships for Data Collection and Sharing (Decade Project 83)
OneArgo: an integrated global, full depth and multidisciplinary ocean observing array for beyond 2020 (Decade Project 114)
GO-SHIP Evolve (Decade Project 3)
Digital Twins of the Ocean - DITTO (Decade Programme 137)

NOTES: See Appendix A, Table A.2, for a description of each. WOC, World Ocean Council.
[a] See https://www.oceandecade.org/decade-actions.

    - enhancing the usability of data and/or models for various applications, including approaches such as digital twins.
- Facilitate partnerships between developers of major ocean data tools to foster coordination and integration by building on existing efforts to create global standards for metadata, query, and data tagging that allow existing data sets to be interconnected, analysis-ready, and automatically accessed.
- Explore the constraints and possibilities regarding high-bandwidth communications from *in situ* ocean data sources.
- Lay out a roadmap for data and computationally intensive cyberinfrastructure that support a digital ecosystem for ocean data, analysis, and simulation.
- Project the expansion in volume over the next decade of observational and simulation data related to the ocean broadly by

  - forging partnerships with ongoing European Union and global data efforts;
  - forging partnerships with major cloud providers; and
  - forging partnerships with commercial ocean, weather, and climate data providers.

It will be important to demonstrate the utility of such data, including economic, social, and scientific values as well as information about the level of accuracy and quantification of uncertainty for use in various applications. For example, there could be an accuracy indicator for data that are informative for understanding and addressing climate change.

Potential Next Steps

Following the initial workshop on An Inclusive and Equitable Ocean, a workshop will be held to bring together a broad spectrum of key public- and private-sector players and other participants in the ocean data space. The outcomes from this foundational workshop will help to guide the establishment and evolution of the data component for each topical theme. Specific goals of these workshops will include the following:

- Develop strategies to improve coordination of domestic and international efforts in the ocean data landscape, particularly around data governance, and the required cyberinfrastructure.
- Develop a pathway toward platforms that enable decision-makers, journalists, the public, and scientists/analysts to readily visualize, analyze, and develop tools and services for holistic approaches to ocean sustainability.
- Identify data strategies and services that support new approaches and capabilities for ocean sustainability and management and ensure openness and engagement with a broad and diverse community. Particular attention should be paid to lowering technical and financial barriers to access data, models, and knowledge services.
- In concert with An Inclusive and Equitable Ocean, incorporate strategies to relate place-based and qualitative knowledge to the more quantitative and large-scale, big data approaches and ways to incorporate different types of data, considering different types of evidence that are important for the co-creation of knowledge.
- Develop potential guidelines for
  - co-designing data policy protocols to support and scale data sharing by partners;
  - building on existing data governance frameworks to create high-level guidelines that inform ongoing efforts to liberate and use ocean data in effective and equitable ways; and

    - developing tools and services to increase the potential application of the data, with an emphasis on user-centered design and knowledge services.
- Develop an understanding of hardware and software requirements to enable an advanced ocean data ecosystem that spans high-performance, cloud, and edge computing and accelerates the flow from data acquisition to data-driven services.
- Identify specific data system needs and obstacles for each topical theme.

Defining Success

The committee's vision is to create a platform, or network of platforms, that enables connectivity and builds trust among developers and users. Unlike previous efforts, the focus is on creating open, shareable, and accessible networks, rather than on the technologies of computation and storage which will inevitably evolve. By the end of the UN Ocean Decade, the committee envisions a network that connects the unconnected, truly engaging those who need knowledge services for ensuring a healthy ocean with those who can create those services. This will create a new era of scalability and flexibility, enabling new solutions to be developed and deployed rapidly rather than relying on the static and inflexible systems of today that are designed primarily to address specific research questions.

TOPICAL THEMES

The Ocean Revealed

Overview of Theme

Understanding fundamental ocean processes (e.g., ocean circulation, the carbon pump, food web dynamics) is the key to developing sustainable approaches to the many human uses of the ocean and its resources. Advancements in scientific understanding have often followed technological advancements in the ability to observe, from telescopes for exploring the universe to electron microscopes for visualizing large molecules and minute biological structures. In Earth sciences, sensors deployed from satellites and airplanes have provided a global perspective through remarkable images of terrestrial and atmospheric processes that have revolutionized our understanding of Earth's processes. While this technology has revealed an increasingly comprehensive picture of the ocean's surface, features beneath the surface cannot be seen because water attenuates the signals detected by traditional remote sensing. To overcome these constraints, some underwater observing systems take advantage of the efficient propagation of sound in water and use acoustic sensors.

Alternatively, scientists deploy *in situ* instrumentation, historically limited to samples and measurements made from ships or moorings and consequently sparse in terms of both coverage and frequency. Although ship-based observations will remain necessary (for reference standards, process exploration, and technical development), innovative autonomous observing systems and the development of new underwater sensors to deploy on global fleets of profiling floats, moored arrays, and coastal observing systems consisting of gliders and moorings could greatly enhance the capacity to observe essential ocean variables[4] for physical oceanography, chemistry, and biology in coastal, continental shelf, and open ocean regions at useful spatial and time scales. For example, the global Argo profiling float program[5] now gives scientists the ability to track critical climate parameters—temperature, salinity, and circulation—through the top 2,000 m of the ocean. The Ocean Revealed poses the following question: what if a global observing system (or system of systems) were designed with the idea of using multiple innovative technologies and modalities to optimize the range of parameters measured, the number of processes studied, and spatial and temporal coverage?

The impetus for this theme comes from the breadth of innovative approaches to ocean observation contained in the Ocean-Shots, each of them constrained in their scope to a limited range of temporal or spatial observations and resolutions but collectively representing the range of scales and processes that are needed to address critical ocean sustainability problems. What if these innovators and domain experts could be brought together to define (or adopt previous definitions of) the essential parameters that need to be measured and challenged to design a system of systems that could make the needed observations at appropriate spatial and temporal scales? For example, several Ocean-Shots proposed the use of existing undersea cable infrastructure to provide power, communications, and potentially positioning for sensors deployed on the cable. This would promote the development and deployment of innovative autonomous sensors that could roam freely and return to the cables for data transfers and access to power for recharging.

The key to The Ocean Revealed will be to engage a broad spectrum of experts and users such as engineers, acousticians, geologists, geodesists, geophysicists, modelers, biologists, resource managers, chemists, and physical oceanographers with experts in instrumentation, computing, AI/ML, signal processing, information theory, data visualization, and underwater communications. Together, they can collectively design the sensor suites and sampling characteristics needed to capture key ocean processes to address the most important issues of ocean sustainability and to provide an ongoing picture of the changing state of the ocean. Construction of a comprehensive understanding of the ocean on, above, and below the surface will only be possible if these technologies and approaches are developed and deployed in a coordinated fashion that is closely coupled to

[4] See https://www.goosocean.org/index.php?option=com_content&view=article&id=283&Itemid=441.

[5] See https://argo.ucsd.edu.

applications and, in the context of the UN Ocean Decade, focused on supporting sustainable development. The Ocean Revealed will focus on the following two major components.

New technologies and approaches. Implementation of acoustic observations and services will enable the capture of both a synoptic and a long-term picture of ocean processes. In contrast to electromagnetic waves, which may penetrate underwater a few hundred meters at best, sound can propagate for hundreds of thousands of meters and thus can serve as our observational eyes and ears. Because of the pivotal role of sound in our understanding of physical acoustics in the ocean, a myriad of both passive and active acoustic sensors and sensing systems have been developed. However, these sensors have often been deployed locally without a national or global overarching framework. A coordinated acoustic observation network, leveraging existing undersea infrastructure, would provide the ability to conduct ubiquitous sensing of critical ocean variables with a mapped seafloor that provides a complete geospatial context. These observations would provide the opportunity to derive transformative insights into key ocean and seafloor processes that are essential for maintaining a sustainable ocean. Biologists have begun employing acoustic methods to study aggregations of minute plankton, monitor fish populations, and conduct behavioral studies of whales—a species that uses sound to communicate over large distances. Additionally, a global acoustic network tied to existing cable infrastructure offers transformative possibilities in terms of underwater communications and services, increasing the potential for the development of an acoustic localization network (e.g., underwater global positioning system).

Also highlighted in the Ocean-Shots was the transformative potential of environmental deoxyribonucleic acid (eDNA), and for marine microbes, other biomolecular "omic" techniques such as transcriptomics, proteomics, and metabolomics (Samuel et al., 2021). The development of eDNA techniques offers another example of an innovative technology with the potential to bring broad new insights into ocean biology. DNA isolated from a small volume of seawater (i.e., eDNA) reveals the presence of millions of microbes and many hundreds of eukaryotic species. These samples can reveal the presence of otherwise undetected species and provide an indication of their abundance as well as the biological activity of microbes. Further development of this technology spans applications such as fish stock assessment and detection of invasive species to new approaches for monitoring marine protected areas and restoration projects, rapidly assessing changing microbial responses and potential impacts of oil spills, and uncovering the habitats and habits of rare or stealthy species. As a sensitive tool for assessing marine biodiversity, eDNA approaches could help scientists characterize marine ecosystem composition and dynamics that will be critical for future management of marine resources in a rapidly changing environment.

Several of the Ocean-Shots focused on the development of new, low-cost underwater sensors and platforms that, if inexpensive enough, could be deployed in large numbers and offer the opportunity to capture a synoptic picture of some ocean processes. Just how far can these sensors and platforms be pushed? How many key parameters can be measured through inexpensive drifting, wave-powered, or density-driven platforms? With simple, easy to use sensors, can citizen/community science play an important role in extending the coverage of observations? These are some of the questions that will be addressed. The global deployment of inexpensive sensors and citizen/community science, combined with more expensive autonomous platforms and strategically located moored arrays, all with the possibility of communicating with a global cabled infrastructure that may provide power, communications, and positioning, may offer the opportunity to finally capture ocean processes at the scales needed to answer key climate and sustainability questions.

Underexplored ocean regions. Another key component of The Ocean Revealed is a focus on underexplored ocean regions. With only approximately 20 percent of the world's seafloor directly mapped at the resolution achievable by modern surface-ship mapping systems, and far less of the ocean's volume sampled or explored, vast areas of the ocean exist where the morphology and environmental conditions are totally unknown (Mayer et al., 2018). This is particularly true in the deep ocean and mesopelagic zones (below 200 m), which despite containing the largest collection of habitats on the planet remain the least observed.

These dynamic ocean regions provide critical climate regulation, house a wealth of mineral resources, and harbor tremendous biodiversity, much of which remains to be characterized. However, their inaccessibility and vast size has made them difficult to explore. Likewise, the harsh environment of the polar oceans has hindered scientific exploration and understanding. The Arctic is particularly important because it is undergoing rapid changes owing to a warming climate and plays a key role in the response of the global climate system. Both the Arctic and the Antarctic contain vast ice sheets whose rapid melt and/or collapse have become the largest contributor to global sea level rise.

The processes driving these changes are inextricably linked to their adjacent oceans. With an expanding suite of sensors capable of measuring a range of critical ocean variables, autonomous vessels are now opening up these regions to study, but field programs are often short term, seasonally biased toward summer, and locally focused. Numerous Ocean-Shots highlighted these underexplored regions and suggested observational systems and approaches required to address the critical problems associated with them. Including experts from these regions in discussion of the observational system of systems will assure that the special challenges associated with these regions will be addressed.

Although exploration is generally defined in terms of physical place (e.g., the deep seafloor), it also refers to the time and space scales of the ocean and oceanic processes. The time scales of ocean ecosystems are often a day or less, yet a satellite might be able to observe them only every few days. Variations of the Atlantic Meridional Overturning Circulation (AMOC) happen on weekly to century time scales, but sustained observations are only a few decades old, with observations only now starting in some regions. And many critical areas of the ocean (e.g., the Southern Ocean and Arctic) are only sparsely observed except by satellites and small but growing numbers of under-ice profilers. Yet, new technologies again show promise to extend observing systems to sense the ocean at the required time and space scales—this too will be a key component of design.

Through thoughtful and coordinated design and application, these and other new technologies will open up the deep ocean, the Arctic and Southern Oceans, and other "dark" ocean regions and parameters to new levels of study and understanding. By illuminating the properties of the ocean beneath the surface, researchers will generate a better understanding of how the ocean supports human communities and influences the critical properties of our planet. This fundamental information forms the basis for the development of sustainable practices and will facilitate achievement of the goals of other topical themes.

The Ocean Revealed calls for the coordinated development and deployment of new techniques, technologies, and approaches, including citizen/community science where possible, to measure a range of ocean variables (biological, chemical, and physical) in understudied ocean regions to provide the critical understanding of ocean processes needed for sustainable development. It builds on the potential of leveraging existing cable infrastructure and developing an enhanced understanding of the ocean's acoustic environment and the rapid development of eDNA as a remote sensor of ecosystem dynamics. Processes that could be studied with these new technologies include heat, mass, and biogeochemical tracer transport; circulation and mixing; ecosystem health; and marine organism biodiversity and behavior. Each of these have relevance for many environmental priorities such as climate change, ecosystem-based management, commercial and recreational activities, marine hazards, and maritime safety.

Decade Challenges and Outcomes Addressed

This theme addresses a number of UN Ocean Decade challenges including Challenge 7: "Expand the Global Ocean Observing System - Ensure a sustainable ocean observing system across all ocean basins that delivers accessible, timely, and actionable data and information to all users" and Challenge 8: "Create a digital representation of the Ocean - Through multi-stakeholder collaboration, develop a comprehensive digital representation of the ocean, including a dynamic ocean map, which provides free and open access for exploring, discovering, and visualizing past, current, and future ocean conditions in a manner relevant to diverse stakeholders."[6] This theme also addresses many of the UN Ocean Decade

[6] See https://www.oceandecade.org/challenges.

outcomes, most directly Outcome 4: "A predicted ocean where society understands and can respond to changing ocean conditions." It also addresses Outcome 2: "A healthy and resilient ocean where marine ecosystems are understood, protected, restored and managed"; Outcome 3: "A productive ocean supporting sustainable food supply and a sustainable ocean economy"; and Outcome 5: "A safe ocean where life and livelihoods are protected from ocean-related hazards" (UNESCO-IOC, 2021a).

Connections to Ocean-Shots, UN Ocean Decade Actions, and U.S. Ocean Priorities

The Ocean Revealed builds on more than 30 Ocean-Shot concepts and a number of UN Ocean Decade–endorsed actions listed in Table 2.3. This theme is further supported by other decade-related resources including NSTC's U.S. Decadal Vision for America's oceans, which underscores the importance of new technologies including acoustic measurements for exploration, discovery, and long-term monitoring and advancement of the blue economy, and highlights the importance of understanding the changing conditions in the Arctic.

Potential Research Elements

The coordinated development and application of new and emerging sensing techniques and technologies that take advantage of autonomous systems, which leverage ML and AI to map, characterize, and understand the underwater environment, includes the following:

- Development of inexpensive sensors and platforms for ubiquitous deployment.
- Monitoring of vast ocean areas through the development and deployment of passive and active acoustic sensors.
- The complete mapping of the seafloor by the end of the UN Ocean Decade to provide a geospatial context for all other measurements.
- The development of underwater location and communications services, accessing power and communication services through existing submarine telecommunications cable infrastructure. This infrastructure would support the expanded deployment of biogeochemical, under-ice, and full ocean depth autonomous sensors worldwide; active acoustic observations of biomass volumes, marine organism distribution, and behavior; mass, heat, and biogeochemical tracer transport and mixing; and seafloor topography. These observations can be made at a range of scales from microstructure to basinwide.

TABLE 2.3 Connections of The Ocean Revealed to Ocean-Shots and UN Ocean Decade Actions

Title
Ocean-Shots
Long-Term, Global Seafloor Seismic, Acoustic and Geodetic Network
Unlocking the secrets of the evolving Central Arctic Ocean Ecosystem: A foundation for successful conservation and management
Arctic Shelves: Critical Environments in Flux
Integrated Ocean Observing Across the Northwest Atlantic
Ocean Arc: An Ocean Shot for the Arctic
Challenger150: A Global Survey of Deep Sea Ecosystems to Inform Sustainable Management
DORIS: Deep Ocean Research International Station
Observing the Oceans Acoustically
Ocean Sound Atlas
Auscultating the Oceans: Developing a Marine Stethoscope
Implementing a Global Deep Ocean Observing Strategy (iDOOS)
SMART Subsea Cables for Observing the Ocean and Earth
Measuring the Pulse of Earth's Global Ocean
The Endless Dive: Marine Species 3D response to climate change in oceans
Accelerating Global Ocean Observing: Monitoring Coastal Ocean Through Broadly Accessible, Low-Cost Sensor Networks
Great Global Fish Count by DNA
A Global eDNA Monitoring System (GeMS)
The US Ocean Biocode
Boundary Ocean Observation Network for the Global South (BOON-GS)
Measuring the Ocean: A Plan for Open Source Underwater Robots and Sensors to make Ocean-Science more Accessible
Ocean-Shots
METEOR: A Mobile (Portable) Ocean Robotic Observatory
Sustained, Open Access, In-situ, Global Wave Observations for Science and Society
Measuring Global Mean Sea Level Changes With Surface Drifting Buoys
Super Sites for Advancing Understanding of the Oceanic and Atmospheric Boundary Layers
Twilight Zone Observation Network: A Distributed Observation Network for Sustained, Real-time Interrogation of the Ocean's Twilight Zone
Persistent Mobile Ocean Observing: Marine Vehicle Highways
FathomNet: Exploring Our Ocean Using Artificial Intelligence
Battery-free Ocean Internet-of-Thing (IoT)
Improved Value of the Observing System through Integrated Satellite and in situ Design

TABLE 2.3 Continued

Title
A Real-Time Global Rivers Observatory
Butterfly: Revealing the Ocean's Impact on Our Weather and Climate
A Global Network of Surface Platforms for the Observing Air-Sea Interactions Strategy (OASIS)
Southern Ocean Storms - Zephyr
UN Ocean Decade Endorsed Actions[a]
Ocean Biomolecular Observing Network (OBON) (Decade Programme 26)
OneArgo: An Integrated Global, Full Depth and Multidisciplinary Ocean Observing Array for Beyond 2020 (Decade Project 114)
Ocean Decade Research Programme on the Maritime Acoustic Environment (UN-MAE) (Decade Programme 12)
IOGP Environmental Genomics Joint Industry Programme (Decade Contribution 1)
International Ocean Discovery Program (Decade Contribution 140)
IOGP Sound and Marine Life (SML) Joint Industry Programme (JIP) (Decade Contribution 42)
Deep Ocean Observing Strategy (Decade Programme 129)
The Nippon Foundation-GEBCO Seabed 2030 Project (Decade Programme 107)
Promote Seabed 2030 and Ocean Mapping (Decade Contribution 133)
Global Ocean Biogeochemistry Array (GO-BGC Array) (Decade Contribution 142)

NOTES: See Appendix A, Table A.3, for a description of each. GEBCO, General Bathymetric Chart of the Oceans; IOGP, International Association of Oil & Gas Producers.
[a] See https://www.oceandecade.org/decade-actions.

- Capture of global biodiversity and fisheries and marine resources through internationally coordinated programs of eDNA sampling that may also involve citizen/community science.
- A coherent plan for combining and fusing data across subdisciplines for comprehensive ocean sensing and monitoring for hazards such as earthquakes and submarine landslides, which may trigger tsunamis; climate indicators such as the strength of the AMOC, acoustic and eDNA techniques for monitoring exploited fish populations, and biodiversity "baselines"; biogeochemical sensors to detect and monitor ocean acidification and deoxygenation; and development of new sensors and techniques for monitoring illegal, unreported, and unregulated fishing, plastic pollution, and impacts of future seabed mining activities.
- Bioprospecting for new or understudied species possessing medicinally beneficial compounds.

Potential Next Steps

Using the submitted Ocean-Shots as a starting point, a workshop will be held to bring together groups interested in technology development (sensors and platforms), observations, and applications for sustainable development. The participants will be asked to focus on the "system of systems" that will provide sufficient coverage to reveal ocean properties and processes from the surface to the depths across the world ocean. The goals of the workshop will include the following:

- Outline the overarching components and structure of the envisioned system for providing the desired end-products, including the essential ocean variables that need to be measured and monitored;
- Initiate the development of a rigorous architecture to establish a strategy for investment and deployment, assuring interoperability across systems and sensors and the optimization of platforms for multiple parameter data collection; and
- Ensure the involvement of the broader community in the development, implementation, and sharing of data from proposed systems.

The key to the success of these workshops will be to include representation from communities beyond those represented in the submitted Ocean-Shots (e.g., subject-matter experts in signal processing, AI/ML, telemetry, modeling and visualization, ocean state estimation, and bioinformatic taxonomy;[7] cable and electrical engineers; and the users of the data and data products) so that even in this initial planning stage, all aspects of the theme can be discussed, including developments outside of the marine realm. Opportunities to link undersea networks with above-sea systems, including satellite networks, will be explored. Representatives of the foundational themes will participate to ensure that their findings will be reflected in planning next steps.

Defining Success

By 2030, The Ocean Revealed would provide the observations necessary to support the Sustainable Development Goals and growth of the blue economy. As part of this initiative, observation needs will be identified through community engagement. This would include globally coordinated observations of key ocean variables through the combination of newly developed inexpensive sensors, eDNA, and other "omics" techniques that can be broadly deployed; strategically located moorings measuring multiple parameters; and enhanced autonomous sensing capabilities aided by new developments in AI/ML and high-bandwidth satellite communications. These will be supported by a global array of active and

[7] Use of DNA sequence homology to construct a taxonomic tree.

passive acoustic sensors and take advantage of existing cabled infrastructure to provide power, communications, and positioning.

Such an observation "system of systems" would enable the following:

- Climate services to enhance prediction (reduce uncertainties) and to enable mitigation of and adaptation to hazards (e.g., sea level rise, marine heatwaves, and other climatic events);
- Sub-seasonal to seasonal weather forecasts based on enhanced understanding of ocean and atmospheric coupling;
- More accurate forecasting and early warning systems to address maritime and coastal hazards and safety, including water level elevation, as a consequence of hurricanes and tsunamis;
- Long-term monitoring of ocean hydrographic (thermal and saline) structure and sensitivity of marine organisms to hydrographic changes;
- Conservation of biodiversity through detection and monitoring of species; and
- Ecosystem-based fisheries management through enhanced monitoring of ecosystem components in addition to the target species.

The Restored and Sustainable Ocean

Overview of Theme

The ocean is subject to many pressures from human activities that change ocean ecosystems. Examples of such pressures include the direct and indirect effects of fishing; alteration of habitat from shoreline and port development, navigational dredging, sand, and other mineral extraction; pollution by terrestrial run-off of chemical and plastic waste; and the impacts of carbon dioxide emissions, which cause ocean acidification, elevated water temperatures, and deoxygenation. The impacts of these pressures vary across ocean ecosystems, disproportionately affecting coral reefs, coastal marshes, and the rapidly warming and melting Arctic.

Human communities rely on the many ecosystem services provided by the ocean and with human populations increasing, the dependence on ocean resources is likely to increase, in particular for renewable energy, deep-sea minerals, and both wild caught and farmed seafood. Recognizing the growing demand for marine resources, the international community has promoted policies to protect and sustainably use the biodiversity of areas beyond national jurisdiction, including the deep sea (IOC-UNESCO, 2020). Ensuring that the ocean will continue to meet these needs will require both restoration of what has been lost or degraded and sustainable management of ocean biodiversity and productivity to support food production and other uses, such as marine bioprospecting for novel compounds effective in treating human diseases. To achieve the UN Ocean Decade outcomes of a healthy, resilient, and productive ocean; the U.S. Decadal Vision of promoting economic prosperity and safeguarding human health; and the High Level Panel for a Sustainable Ocean Economy goals of protecting effectively, producing sustainably, and prospering equitably requires both restoration and effective management (Stuchtey et al., 2020). Meeting these goals is made more difficult by the fact that the changing Earth climate system is altering the composition and distribution of marine ecosystems such that ocean ecosystems are "on the move" with consequent changes in productivity. Although in some cases reducing the activity causing the decline is sufficient for recovery, in other situations active interventions will be required to restore habitats and ecosystems and ensure their resilience.

Restoration of coastal systems, such as mangroves, seagrasses, coastal marshes, coral reefs, and marginal sea ice, is a high priority because they provide a number of key ecosystem services such as nursery habitat for a variety of fishes and protection from storm surges. Due to human activities, these habitats have experienced devastating loss and degradation in some areas and a science-based restoration approach will be required to recover these valuable ecosystem services. Coral reefs are a high-profile example of a marine ecosystem requiring both restoration and protection. Coral reefs provide a suite of benefits to society through coastal protection (buffering of wave energy and storm surge), productive fisheries, tourism and recreation, and high levels of biodiversity. Many stressors affect coral reefs, with a warming climate posing the most existential threat (NASEM, 2019). New approaches for enhancing the capacity of corals to survive in a warmer world in concert with a reduction in other stressors, such as pollutants and unsustainable fishing practices, promise a pathway toward maintaining these ecosystems for future generations.

Sustainable food production is a critical ecosystem service provided by ocean ecosystems. The ocean supports commercial and subsistence fishing, mariculture, and even terrestrial livestock production. The ocean is vital in meeting the nutritional needs of many Indigenous and disadvantaged communities who often rely on locally sourced foods. Ocean food production includes three distinct areas: commercial fisheries, mariculture, and feedstocks for aquaculture and agriculture.

Commercial fisheries production. The cumulative impact of changing environmental conditions; failures in resource management by coastal nations; and illegal, unreported, and unregulated fishing in international waters threatens to reduce the contribution of wild-caught fisheries to global food production and security. Additionally, with changing ocean productivity and shifts in distribution, the equilibrium assumptions that have underpinned fisheries management are now in question and a comprehensive re-evaluation is warranted. A particular challenge is to improve our understanding and deployment of approaches to realize the complex social-ecological systems that characterize fisheries. Understanding the intricacies of fisheries management will require new partnerships and new frameworks.

Mariculture/aquaculture production. Aquaculture is the fastest growing sector in global food production, predicted to increase by 32 percent over 2018 production levels by 2030 (FAO, 2020). In the United States, the federal government has identified two initial Aquaculture Opportunity Zones (Southern California and the Gulf of Mexico) as a means to incentivize investments in the sector; at the same time, states have taken a mixed approach with respect to permitting operations. Innovations in aquaculture, including a focus on new species for culture, bioengineering of target species, and technical innovations, such as recirculating systems on land, are opening new opportunities to support consumption while minimizing impacts on ocean ecosystems.

Ocean-sourced feedstocks for aquaculture and agriculture. The majority of landings of forage fish (small pelagics) are reduced into fishmeal and fish oil (Alder et al., 2008) used as high-value nutritional additives for animal feeds, mostly for fish aquaculture, followed by pigs and poultry (Tacon and Metian, 2008). At the same time, recent developments are demonstrating the potential for plant-based feedstocks (including seaweed and algae) to reduce pressure on forage fish stocks and to reduce the carbon footprint of land-based protein production. A detailed understanding of the impacts and ramifications of these activities is essential to ensuring the long-term sustainability of food production. The Restored and Sustainable Ocean will take a "whole-ocean approach" and look at the connectivity among research initiatives focused on the restoration and sustainability of component ecosystems such as coral reefs and other coastal habitats, the deep sea, Arctic and Southern Oceans, and the mesopelagic zone. Particular attention will be on research that incorporates the human ecosystem into new strategies and approaches for ocean resources, health, marine transportation, renewable energy, and resilience.

Decade Challenges and Outcomes Addressed

The Restored and Sustainable Ocean addresses many of the UN Ocean Decade challenges and outcomes. The primary UN Ocean Decade challenges include Challenge 1: "Understand and beat marine pollution - Understand and map land- and sea-based sources of pollutants and contaminants and their potential impacts on human health and ocean ecosystems, and develop solutions to remove or mitigate them"; Challenge 2: "Protect and restore ecosystems and biodiversity - Understand the effects of multiple stressors on ocean ecosystems and develop solutions to monitor, protect, manage and restore ecosystems and their biodiversity under changing environmental, social and climate conditions"; and Challenge 3: "Sustainably feed the global population - Generate knowledge, support innovation and develop solutions to optimize the role of the ocean in sustainably feeding the world's population under changing environmental, social and climate conditions."[8] This theme is relevant to Outcome 1: "A clean ocean where sources of pollution are identified and reduced or removed"; Outcome 2: "A healthy and resilient ocean where marine ecosystems are understood, protected, restored and managed"; and Outcome 3: "A productive ocean supporting sustainable food supply and a sustainable ocean economy." However, the theme also addresses other outcomes, including Outcome 4: "A predicted ocean where society understands and can respond to changing ocean conditions" and Outcome 7: "An inspiring and engaging ocean where society understands and values the ocean in relation to human well-being and sustainable development" (UNESCO-IOC, 2021a). The ocean delivers an enormous array of services to the blue economy, and its health is invaluable to the planet. Moreover, the beauty and wonder of the ocean provide a cultural resource that inspires the arts and offers respite for many who enjoy beaches, swimming, diving, fishing, and other water activities.

Connections to Ocean-Shots, UN Ocean Decade Actions, and U.S. Ocean Priorities

The Restored and Sustainable Ocean encompasses ideas from approximately 20 Ocean-Shots (see Figure 2.2) and complements many of the UN-endorsed actions (see Table 2.4). It is consistent with several priorities of NSTC's U.S. Decadal Vision (SOST, 2018). A selection of some of the most relevant Ocean-Shots and UN Ocean Decade activities is included in Table 2.4.

Potential Research Elements

- Utilize observing tools and systems from The Ocean Revealed to measure ecological processes and enable the linking of ecological and physical models with socioeconomic models;
- Enable the exploration of "what if" scenarios to understand natural and technological risks and uncertainties with different policy interventions, including the role of marine spatial planning as a means to reduce these risks, for both developed and developing countries;
- Scale up new nature-based solutions (NBSs) to improve health and increase resilience of ocean ecosystems; and
- Investigate the cumulative impacts of land-based pollution, including plastic waste, microplastics, and fertilizer run-off, on ecosystem health and resilience of urban seas.

[8] See https://www.oceandecade.org/challenges.

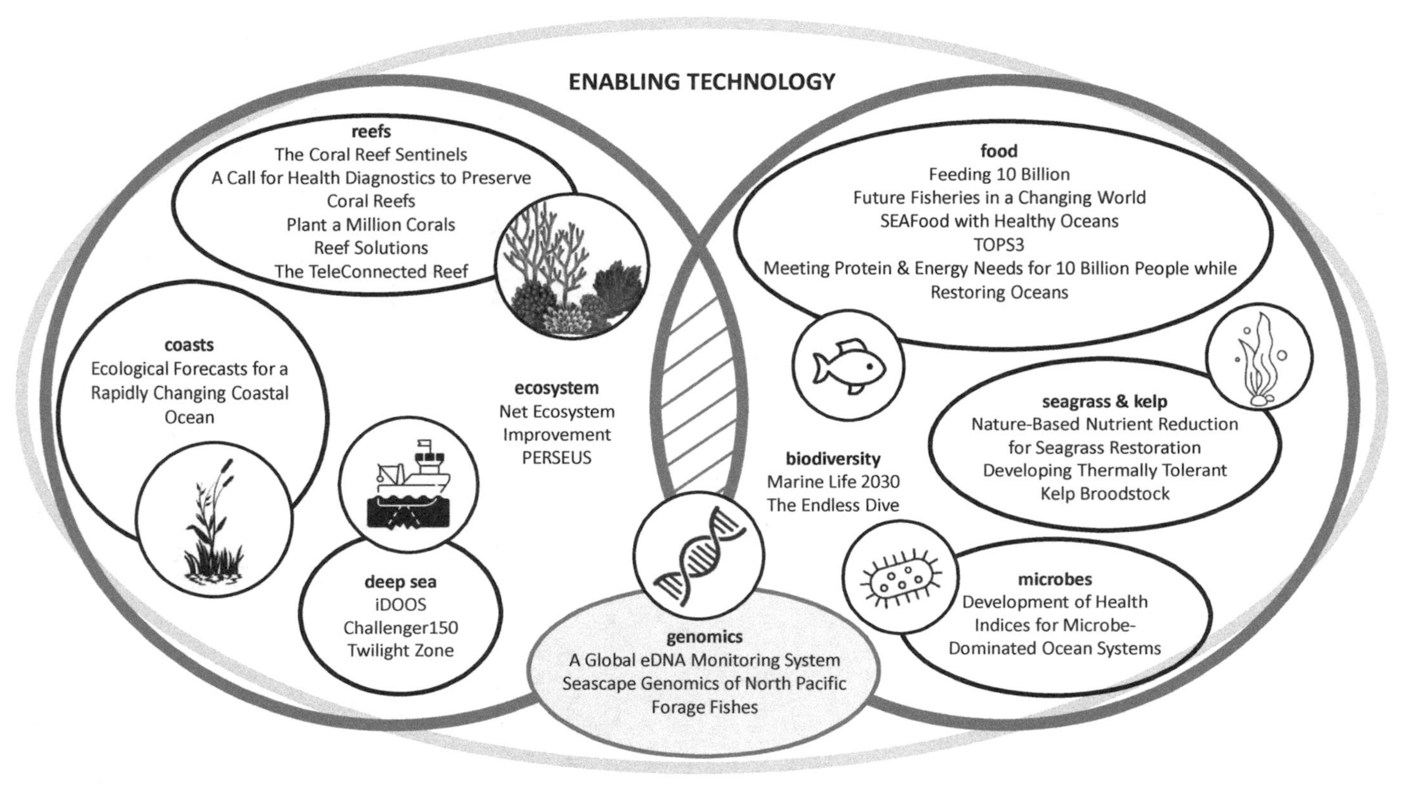

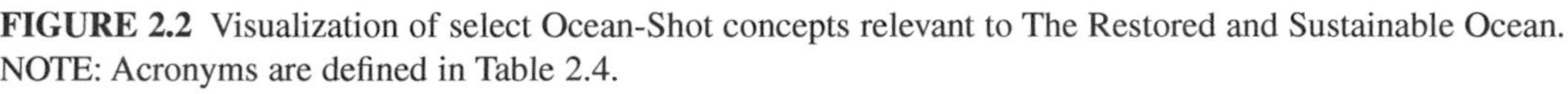
FIGURE 2.2 Visualization of select Ocean-Shot concepts relevant to The Restored and Sustainable Ocean.
NOTE: Acronyms are defined in Table 2.4.

TABLE 2.4 Connections of The Restored and Sustainable Ocean to Ocean-Shots and UN Ocean Decade Actions

Title
Ocean-Shots
The Coral Reef Sentinels: A Mars Shot for Blue Planetary Health
Implementing a Global Deep Ocean Observing Strategy (iDOOS)
Challenger150: A Global Survey of Deep Sea Ecosystems to Inform Sustainable Management
Twilight Zone Observation Network: A Distributed Observation Network for Sustained, Real-time Interrogation of the Ocean's Twilight Zone
Marine Life 2030: Forecasting Changes to Ocean Biodiversity to Inform Decision-Making - A Critical Role for the Marine Biodiversity Observation Network (MBON)
Ecological Forecasts for a Rapidly Changing Coastal Ocean
Net Ecosystem Improvement: An Evidence-Based Approach
Feeding 10 Billion: Contributions from a Marine Circular Bioeconomy
Future Fisheries in a Changing World
Science Enables Abundant Food (SEAFood) with Healthy Oceans
Transforming Ocean Predictions for Seafood Security and Sustainability (TOPS3)
Meeting Protein & Energy Needs for 10 Billion People While Restoring Oceans
The Endless Dive: Marine Species 3D response to climate change in oceans
A Call for Health Diagnostics to Preserve Coral Reefs
Plant a Million Corals
Reef Solutions: Convergence of Research and Technology to Restore Coral Reefs
The TeleConnected Reef
Seascape Genomics of North Pacific Forage Fishes
Developing Thermally Tolerant Kelp Broodstock to ensure the Global Persistence of Kelp Mariculture in Response to Ocean Change
PERSEUS (Pelagic Ecosystem Research: Structure, Emergent Functions, and Synergies)
Development of Health Indices for Microbe-Dominated Ocean Systems
Nature-Based Nutrient Reduction for Seagrass Restoration
UN Ocean Decade Endorsed Actions[a]
Deep Ocean Observing Strategy (DOOS) (Decade Programme 129)
Fisheries Strategies for Changing Oceans and Resilient Ecosystems by 2030 (Fish-SCORE 2030) (Decade Programme 63)
Global Ecosystem for Ocean Solutions (GEOS) (Decade Programme 172)
Ocean Biomolecular Observing Network (OBON) (Decade Programme 26)
Sustainability of Marine Ecosystems through global knowledge networks (SMARTNET) (Decade Programme 90)
NOAA Coastal Aquaculture Siting and Sustainability Program (Decade Contribution 51)

TABLE 2.4 Continued

Title
NSF Coastlines and People (Decade Contribution 135)
Sustainability, Predictability and Resilience of Marine Ecosystems (SUPREME) (Decade Programme 118)
Coral Reef Restoration Engaging Local Stakeholders Using Novel Biomimicking IntelliReefs (Decade Project 112)

NOTES: See Appendix A, Table A.4, for a description of each. NOAA, National Oceanic and Atmospheric Administration; NSF, National Science Foundation.
[a] See https://www.oceandecade.org/decade-actions.

More specific research elements are organized under the following categories.

- Restoration of critical habitat:
  - adopt new technological applications and approaches to restore lost habitat that is critical to sustain certain ecosystems (e.g., coral restoration, new approaches to loss of sea ice and mitigation of consequences for charismatic megafauna, mangrove replanting/recovery);
  - develop advanced propagation and transplantation techniques for key species required for restoring habitat and depleted species; and
  - develop and test novel, low-cost approaches for monitoring vulnerable habitats and restoration projects for identifying problems and developing solutions.
- Food security:
  - address the challenges of developing a sustainable mariculture industry that improves the livelihoods and resilience of coastal communities without harming the environment,
  - address the challenges of meeting the growing demands for marine resources in aquaculture and livestock operations,
  - optimize marine transportation assets and shipping routes to connect sources with needs for food supply and minimize impacts on local food sources, and
  - develop and assess ecosystem-based approaches to fisheries management that can be considered within a broader ecosystem-based approach and that are resilient to climate change.
- Nature-based solutions:
  - explore an implementation framework for the 2022 National Academies report *A Research Strategy for Ocean-Based Carbon Dioxide Removal and Sequestration*;
  - develop lessons learned from existing efforts to design resilient coastal protection that are mainly comprised of NBSs;

  - enhance beneficial uses of dredge materials to provide a win-win for maritime transportation and ecosystem restoration; and
  - measure the ecological, economic, and social benefits from Marine Protected Areas to better ensure that gains are equitably distributed and resources are protected effectively.
- Ocean energy production:
  - characterize the diversity and dynamics of seafloor ecosystems globally but with an initial focus on areas targeted for deep-sea mining and offshore energy development (e.g., wind turbines and hydrokinetic energy facilities),
  - estimate recovery rates for a variety of ecosystem components from disturbance by ocean energy production activities, and
  - increase understanding of the connectivity of the deep sea with mesopelagic and surface ecosystems.

Potential Next Steps

The workshop for this theme should include separate sessions on each subtheme: coastal systems, food security, and ocean biodiversity. This would be followed by an integrative session to identify commonalities across these topics. The goal of the workshop will be to identify research opportunities that will help design and evaluate solutions that meet the three goals: to protect effectively, produce sustainably, and prosper equitably. Potential topics could include the following:

- Linking of new classes of ecosystem measurements to coupled natural–human management models to better inform ecosystem-based management;
- Incorporation of Indigenous knowledge into the restoration and protection of coastal ecosystems and into the fishing and aquaculture processes that rely on these ecosystems;
- Diversification of participation, target species, and production systems in the aquaculture industry;
- Adaptive management of social-ecological systems that enables both ecosystem and economic services in a changing ocean to ensure resilient coastal communities;
- Identification of educational opportunities to promote ocean literacy and equitable access to ocean science and resources;
- Demonstration projects for ecosystem-based resource management and habitat restoration to reduce coastal flood impacts; and
- Strategies for fishery management that incorporate climate-driven environmental change to foster sustainable practices and food security.

The workshop will include the relevant Ocean-Shot authors, resource managers, and other experts to determine (1) how they can link their efforts; (2) where overlaps may exist and programs could be aggregated; (3) what enabling technologies (both traditional and nontraditional) will be needed to address the problem; (4) how to engage local communities; and (5) most critically, how to identify gaps that would prevent them from achieving the desired outcome and approaches to filling these gaps. Subsequent meetings, based on the interests of the participants, will be encouraged to further evolve the concept.

The ultimate objective of the concluding integrative session will be to describe the desired end state (i.e., new approaches to ecosystem health and resilience) and then develop a bold and transformative program to achieve this end state. Representatives of the foundational themes will participate in the development and execution of the workshop to assist in incorporating their findings into the workshop design.

Defining Success

By the end of the decade, co-production of social-ecological system knowledge will have yielded place-based approaches to more effectively restore and protect vulnerable coastal ecosystems and thus ensure the provision of ecosystems valued in coastal regions. Scientific progress will support implementation of evidence-based practices to restore and protect marine habitats with documented progress by the end of the decade. These practices will recognize and empower Indigenous voices and learn from and support traditional knowledge of coastal ecosystems for the benefit of all. Included in the vision are the following:

- The transdisciplinary science necessary to develop a fuller understanding of the social-ecological system that supports fisheries and ocean aquaculture.
- The development of robust climate-proof ecosystem-based approaches that will be routinely used in commercial and recreational fisheries management.
- A dramatically improved understanding of deep-sea and seabed ecosystems globally, with an initial focus on areas targeted for ocean energy production and seafloor mining for critical minerals. This will address the rapidly growing interest in mineral exploitation in the face of substantial challenges of conducting comprehensive baseline assessments of seafloor ecosystems (e.g., Drazen et al., 2020).
- A thriving ocean aquaculture industry that will contribute significantly to the economies of coastal communities and provide ecologically sustainable yields of marine consumables including seaweeds, shellfish, and finfish.

Ocean Solutions for Climate Resilience

Overview of Theme

The ocean is the major regulator of Earth's climate. The climate's response to increasing levels of greenhouse gases, in particular carbon dioxide, is fundamentally dependent on the ocean and profoundly impacts the ocean. The ocean absorbs about 90 percent of Earth's excess heat and close to 30 percent of anthropogenic carbon dioxide, buffering the effects of greenhouse gas emissions that otherwise would have caused dramatic climatic warming (NASEM, 2017). However, the ocean's capacity is not infinite, and a decrease in buffering will have severe consequences for both the ocean and future climate conditions. Adverse impacts of emissions on the ocean manifest in a number of ways and on a wide range of time scales, including ocean acidification; deoxygenation; shifts in circulation; strengthening of extreme events fueled by oceanic conditions, such as marine heat waves or hurricanes; and sea level rise.

These changes, in turn, affect marine ecosystems either directly (e.g., coral reef bleaching, migration of fish populations) or indirectly through a complex interdependence among ocean circulation, biogeochemical cycles, and ocean ecosystems. All of the themes have climate components, several of which are directly addressed in the other themes identified by the committee. In this theme, two climate resilience topics are highlighted because of their potential for transformative change, while recognizing that other climate topics, such as warming waters and ocean acidification, are also critical for achieving a sustainable ocean.

Coastal resilience. Rising seas and more severe storms threaten coastal and island communities globally, with dire consequences for those without the resources to adapt. Waterfronts serve tourism, recreation, energy industries (oil and gas, offshore wind), fisheries, shipping, and naval operations. For example, more than 50 percent of oil refineries in the United States are located on the coast (Thatcher et al., 2013), along with natural gas facilities and major transportation corridors.

Higher water levels, driven by melting ice sheets and thermal expansion, cause land loss from both inundation and erosion, raise the frequency of tidal (sunny day) flooding, and increase the vulnerability of coastal regions to flooding from severe storms. Conventional strategies for protecting coastal zones (e.g., seawalls, revetments, groins, and bulkheads) often result in a loss of the natural habitats that support coastal ecosystems and provide the amenities for tourism and recreation (e.g., sandy beaches, recreational fisheries, bird sanctuaries, and nature reserves) (NRC, 2007). New strategies for coastal resilience, such as NBSs, will be required to sustain the range of activities and valued resources found in the coastal zone.

By applying social and human behavioral science studies of risk communication, researchers will be better able to work with coastal communities and

Indigenous nations to collectively identify population vulnerabilities and information needs. Through this cooperative approach, improved predictions of weather and climate will be more integrated into community planning for, and response to, more frequent extreme events like flooding, heat, or drought. Communities will benefit from new global storm-resolving models and decision-support systems to anticipate and effectively react to flooding from severe storms. These Earth system models will require incorporation of ocean processes and observations through advanced data assimilation to improve prediction capabilities. Improved multi-hazard risk communication will require baseline and event-based data collection on social, cultural, and institutional factors influencing how a range of communities use and interpret forecast information.

Climate mitigation. With the broad international agreement[9] that avoidance of severe climate impacts will require limiting global warming to no more than 1.5°C (IPCC, 2018), the challenge is to develop technologies to achieve this goal. This will require first and foremost phasing out fossil fuels and developing renewable energy sources supplemented by active efforts to reduce atmospheric carbon dioxide levels. Offshore wind energy is becoming a mature industry, with hydrokinetic energy conversion mostly in the experimental phase. Decarbonization of the energy infrastructure will increase reliance on rare earth metals for batteries. These metals have been found on the seafloor in some regions, leading to an interest in deep-sea mining.

Ocean-based carbon dioxide removal (CDR) and sequestration strategies have been proposed, ranging from physical, to chemical, to biological approaches for enhancing carbon dioxide uptake in the coastal and open ocean and storage of sequestered carbon dioxide in the seafloor. This is an early, but active, area of research and technology development. The 2022 National Academies report *A Research Strategy for Ocean-Based Carbon Dioxide Removal and Sequestration* provides a detailed assessment of research needs for ocean-based CDR approaches. Importantly, it addresses uncertainties; potential ecological threats; and the social, legal, and ethical contexts associated with each approach. The report lays out a comprehensive research agenda for the coming decade.

It will be equally important to develop sustainable practices for regions hosting new energy development, deep-sea mining, and/or CDR. For this to be informed by the best available science will require comprehensive measurement, quantification, and understanding of "baseline" (i.e., present-day and reconstructed past) conditions, so that impacts can be detected through monitoring. This will also make it possible to anticipate changes caused by extraction of minerals and develop ways in which to sustainably manage or mitigate such changes.

Ocean Solutions for Climate Resilience will examine two components of the extensive intersection between the ocean and climate: (1) the urgent need

[9] See Paris Agreement, https://unfccc.int/sites/default/files/english_paris_agreement.pdf.

to anticipate and plan for coastal change in response to sea level rise and stronger, slower moving, and more frequent coastal storms (e.g., Gori et al., 2022; NASEM, 2018a, 2021, 2022a); and (2) climate mitigation strategies including expansion of renewable energy generation and development of CDR and sequestration approaches. To achieve resilience, climate mitigation and adaptation efforts are required, considering pace, scale, feasibility, and innovation while also addressing the potential environmental impacts and developing sustainable practices for implementation.

Decade Challenges and Outcomes Addressed

Ocean Solutions for Climate Resilience addresses several of the UN Ocean Decade challenges and outcomes including Challenge 5: "Unlock ocean-based solutions to climate change - Enhance understanding of the ocean-climate nexus and generate knowledge and solutions to mitigate, adapt and build resilience to the effects of climate change across all geographies and at all scales, and to improve services including predictions for the ocean, climate and weather"; Challenge 6: "Increase community resilience to ocean hazards - Enhance multi-hazard early warning services for all geophysical, ecological, biological, weather, climate and anthropogenic related ocean and coastal hazards, and mainstream community preparedness and resilience"; Outcome 2: "A healthy and resilient ocean where marine ecosystems are understood, protected, restored and managed"; Outcome 3: "A productive ocean supporting sustainable food supply and a sustainable ocean economy"; and Outcome 4: "A predicted ocean where society understands and can respond to changing ocean conditions" (UNESCO-IOC, 2021a).[10]

Connections to Ocean-Shots, UN Ocean Decade Actions, and U.S. Ocean Priorities

This theme encompasses ideas from more than 10 Ocean-Shots and overlaps with multiple UN Ocean Decade–endorsed actions and NSTC-identified U.S. ocean priorities. A selection of some of the most relevant Ocean-Shots and UN Ocean Decade activities is included in the table below (see Table 2.5).

Potential Research Elements

To address the scientific challenges described in this theme will require inter- and multidisciplinary teams drawn from the natural and social sciences and engineering. This requirement for interdisciplinary research is recognized in many relevant National Academies reports (e.g., NAS et al., 2005; NASEM,

[10] See https://www.oceandecade.org/challenges.

TABLE 2.5 Connections of Ocean Solutions for Climate Resilience to Ocean-Shots and UN Ocean Decade Actions

Title
Ocean-Shots
Butterfly: Revealing the Ocean's Impact on Our Weather and Climate
Southern Ocean Storms - Zephyr
Observing the Oceans Acoustically
SMART Subsea Cables for Observing the Ocean and Earth
Carbon Sequestration via Drilling-Promoted Seawater-Rock Interactions
Caribbean Observatories (CARIBO): Ocean Storminess at the Western Boundary and Its Impacts on Shelf/Slope Environment and Ecosystems
Measuring Global Mean Sea Level Changes With Surface Drifting Buoys
A Global Network of Surface Platforms for the Observing Air-Sea Interactions Strategy (OASIS)
Navigating the Ocean's Role in Carbon Dioxide Removal
A Real-Time Global Rivers Observatory
Super Sites for Advancing Understanding of the Oceanic and Atmospheric Boundary Layers
Mining Five Centuries of Climate and Maritime Weather Data from Historic Records
Why Paleoceanographic Observations are Needed to Improve Future Climate Projections
OceanPredict.US
A Sensor Network for Mixing at the Ocean's Bottom Boundary
UN Ocean Decade Endorsed Actions[a]
Global Ecosystem for Ocean Solutions (GEOS) (Decade Programme 172)
Observing Air-Sea Interactions Strategy (OASIS) (Decade Programme 97)
Blue Climate Initiative - Solutions for People, Ocean, Planet (Decade Programme 138)
A Transformative Decade for the Global Ocean Acidification Observing System (Decade Contribution 116)
NASA Sea Level Change Science Team (Decade Contribution 33)
ForeSea - The Ocean Prediction Capacity of the Future (Decade Programme 28)
CoastPredict - Observing and Predicting the Global Coastal Ocean (Decade Programme 144)
Science Monitoring And Reliable Telecommunications (SMART) Subsea Cables: Observing the Global Ocean for Climate Monitoring and Disaster Risk Reduction (Decade Project 94)

NOTES: See Appendix A, Table A.5, for a description of each. NASA, National Aeronautics and Space Administration.

[a] See https://www.oceandecade.org/decade-actions.

2022a,b) and in the UN Ocean Decade Implementation Plan. Research could be organized under the following three categories.

Mitigation. By 2030, develop sustainable practices for offshore energy development, deep-sea mining for critical minerals used in renewable energy technologies, and the application of ocean-based carbon dioxide reduction and sequestration strategies to meet climate goals. A diversity of research topics would be included:

- Examine "the interactions and trade-offs between ocean CDR, terrestrial CDR, greenhouse gas abatement and mitigation, and climate adaptation, including the potential of mitigation deterrence" (NASEM, 2022a).
- Enhance the ocean's natural ability for carbon capture.
- Leverage natural carbon capture technologies (e.g., coastal blue carbon) and habitat restoration to preserve depleted, and often endangered, ocean species at risk from a variety of environmental stressors, including climate change.
- Promote research into ocean-based renewable energy production (including offshore wind, wave energy, tidal energy, ocean-based solar photovoltaics, ocean current, and ocean thermal energy conversion) that includes assessment of the social and environmental impacts and comprehensive analyses of material flows involved in renewable technologies.
- Develop a comprehensive research agenda for seabed mining that includes monitoring, and analysis to develop a "baseline" understanding of deep-sea biology and ecosystems. This should include an assessment of the mineral resources, technologies for extraction and their environmental impact, and alternatives for energy storage.

Adaptation. By 2030, develop an accurate picture of the coupled natural–human coastal system, with an emphasis on coastal flooding risks and exposure of vulnerable communities, and environmentally sustainable, equitable, and just adaptation strategies for responding to rising seas and increased coastal flooding. This research has a range of ingredients:

- Understand the social-economic and population dynamics of coastal communities (behavioral economics and sociology) as well as the consequences of migration and relocation movements.
- Assess the impact of human interventions (e.g., construction of levees, canals, and sea walls; dredging), as well as natural resource conservation and restoration activities.
- Characterize uncertainties from "far field" sea level rise, including ice sheet mass loss, ocean warming, decadal climate variability and circulation changes, and low-frequency tidal cycles.

- Characterize uncertainties in current-generation coastal flood maps (bathymetry and topography), which will require
  - an improved and unified time-varying geodetic reference system across the ocean-land continuum (NASEM, 2020a);
  - improved flood-plain mapping;
  - recognition of the time-evolving nature of the geomorphology of the coastline and coastal zone; and
  - understanding of landscape evolution (e.g., river sediment deposition, river deltas, barrier islands), and its response to sea level rise and climate change.
- Characterize uncertainty in changes in extreme precipitation events (e.g., slow-moving and increased-intensity hurricanes, atmospheric rivers).
- Overcome the lack of quantification of risks from compounding effects (superposition from different processes).
- Reduce uncertainty in knowledge regarding the most vulnerable communities.
- Improve understanding of coastal infrastructure at risk (e.g., industry, energy sector, military).
- Increase coordination across federal, state, tribal, and local agencies.
- Improve the accuracy of sub-seasonal weather and ocean numerical prediction within an Earth system framework.
- Expand and advance public–private partnerships for decision support.

Sustained monitoring, simulation, and open science. Following the paradigm that you cannot manage (or even understand) what you cannot observe, an important component is the completion and maintenance of a comprehensive observing system, along with formal synthesis frameworks in order to monitor, quantify, detect, understand, and predict human-induced system changes. This addresses information needs in other themes, in particular The Restored and Sustainable Ocean and The Ocean Revealed. An important element will be the integration of Indigenous, local, and social-economic knowledge systems. Together, this observing system would provide a comprehensive basis for devising ways to mitigate, manage, or adapt to anticipated changes.

Potential Next Steps

A workshop with four parallel sessions is proposed to address the various components of this theme:

- Assessment of the benefits, risks, and sustainable scale potential for ocean-based CDR.
- Evaluation of the status and opportunities (technological and economic) of ocean-based renewable energies, both in U.S. waters and worldwide,

including socioeconomic impediments to their implementation. Topics could include technological progress and transferable technology at the global level; environmental trade-offs of renewable energy that are sustainable and maintain a healthy ocean; and feasibility, export market, job creation, and environmental impact for suitable coastal zones around the world.
- Development of the scientific basis, engineering solutions, and social-economic impacts of coastal adaptation, including approaches to respond to sea level rise and coastal flooding. Topics will be informed by relevant National Academies reports on coastal systems with a particular focus on information-sharing with vulnerable (and often low-income) communities in order to co-develop and implement effective adaptation strategies.
- Development of strategies for assessing and closing critical gaps of a global ocean observing system for climate. Topics could focus on the sustained continuation of existing critical components of the ocean component of the Global Climate Observing System and the closing of critical gaps in the Global Ocean Observing System.
- Creation of an Ocean–Climate Partnership or collective impact organization (NASEM, 2017, 2020b) "to increase engagement and coordination of the ocean observation science community with nonprofits, philanthropic organizations, academia, U.S. federal agencies, and the commercial sector" (Weller et al., 2019).

Defining Success

By 2030, the social, cultural, and institutional drivers needed for a climate resilient future will be fully integrated in the scientific data and tools to enable an adaptable and resilient coastal zone. This transdisciplinary approach will do the following:

- Allow for a common frame of reference for the development of (1) flood/inundation maps that meet the variety of needs of state and federal agencies, and (2) coastal flooding projections that will account for multi-hazard risks (e.g., storm surge and heavy rain) and the geomorphic transformation of the coastal zone under a range of climate change scenarios;
- Inform coastal planning that balances the adaptation of the communities and built systems disrupted by rising seas and extreme events with the economic drivers that encourage coastal development;
- Enable engagement of highly vulnerable U.S. coastal communities in the development of mitigation and adaptation strategies tailored to their particular needs;
- Identify opportunities and challenges for ocean renewable energy; and
- Implement co-development of knowledge and solutions for coastal

communities, particularly those at high risk of and low capacity for adaptation.

By 2030, a comprehensive research program will provide a robust understanding of the merits, efficacy, limitations, and environmental risks of ocean-based CDR. The research will provide a more complete understanding of the ethical, legal, and social context for ocean CDR and sequestration approaches. A second desired outcome related to natural carbon sequestration by the ocean is a better understanding of the potential changes in the ocean's capacity for uptake of carbon dioxide under a range of climate change scenarios.

By 2030, ocean renewable energy will provide a substantial fraction of the nation's electricity demand while balancing the interests of other critical ocean values (e.g., marine conservation, shipping, fisheries). This will include maturation of technologies; scalable deployment; cost-effective operations; and mitigation of environmental impacts for offshore wind, ocean thermal energy conversion, and hydrokinetic energy development.

Healthy Urban Seas

Overview of Theme

Semi-enclosed coastal seas bordering urban centers exemplify the complex intersection and interaction of land and sea with the many human-induced changes to the ocean. They are commonly coincident with zones of intense human activity and can be impacted by polluted wastewater, groundwater, and river outflow subject to changes in weather, climate, and land use far upstream. The density of activities surrounding these urban seas (embayments and estuaries, including watershed impacts from the large river catchment regions) can adversely affect air and water quality, natural resources, and public health. Levels of pollutants introduced into marginal seas can be orders of magnitude greater than those introduced to the open oceans (Onink et al., 2021; UNESCO, n.d.).

Coastal areas generally have high sensitivity to climate change including sea level rise and coastal flooding, rising water temperatures, acidification, and coastal storms. These heavily populated urban seas, typically within marginalized communities that are disproportionately impacted by environmental degradation, represent the most intense and complex sites of human–ocean interaction on the planet. Many of these urban seas support extensive wild capture fisheries and emerging mariculture and aquaculture programs that help meet the increasing demand for fresh local seafood. They may be viewed as global ocean "hot spots" within a more broadly distributed planet-wide framework of anthropogenic and climate impacts. Coastal urban areas are intensely populated not just with people but with critical infrastructure. Port facilities, offshore energy, material extraction platforms, stormwater and sewer outlets, near-shore roads and rail, utilities, and communications are examples of the infrastructure that connects communities to the ocean. It is critical to understand the symbiotic relationship among the community, the built environment, and the ocean with investments in more resilient infrastructure.

Focused application of new observing technologies to comprehensively monitor and model these coupled human–ocean ecosystems will reveal rates and patterns of environmental change, as well as document the efficacy of recovery tactics, pollution reduction strategies, and protection efforts. The latter will contribute to improvements in remediation and conservation activities for greater long-term sustainability. A common framework is needed within which the dynamics, restoration trajectories, and management interventions in multiple urban seas could be compared to provide greater insight into the functioning of these complex environments.

The high concentration of anthropogenic activities in and adjacent to these waters creates natural laboratories to study the intense chemical and biological gradients generated by high-density populations in these semi-enclosed marine ecosystems. They are sites of extensive science and data acquisition, such as the long-term studies of the Chesapeake Bay, Puget Sound, and Mississippi River Delta. Furthermore, ports can be partners in identifying low- or no emission solutions and protecting ecosystems. Ships and other vessels provide ready-to-deploy monitoring platforms. They are ideal settings for innovative testing of novel approaches to sensing and modeling of changing ecosystems because of their ease of access and proximity to research facilities. From a computational perspective, these contained basins will allow greater facility for creating a "digital twin"[11] and testing the validity and accuracy of a range of models for simulating the natural, built, and socioeconomic environments, as well as their interactions. Inputs and outputs would be more readily measured and managed because the constrained scale allows the deployment of fixed observatories and sensors on mobile assets. Connected by proximity, interdisciplinary teams of scientific, technical, policy, management, Indigenous, and stakeholder communities could organize to address the multiple challenges facing these densely populated and heavily used ocean areas and co-develop responses.

Healthy Urban Seas will focus on the activities and changes in these constrained and highly populated regions, all of which are magnified by climate change. Careful design of experiments to capture and understand the impacts of human activities will yield information on mechanisms that could be employed for increasing the health and resilience of urban seas. These large population centers also offer an opportunity for community engagement in problems that are directly relevant to their well-being and sustainability. Owing to the proximity to

[11] A digital twin is a computational model that is dynamically updated with observations of the environment such that it provides a virtual representation of the structure, context, and behavior of the physical system. It is designed as a tool to explore potential outcomes of a variety of interventions to inform decisions (see Glossary).

the ocean, ocean-related issues could be brought into school curricula and public events more easily, involve diverse communities, and be brought to the attention of policy makers.

This theme will draw on most aspects of the foundational and other topical themes. It has the potential to greatly increase understanding of the linkages between the natural system and sociocultural change through expanded collection of data within crucial temporal and spatial frameworks. It offers a superb arena in which to develop cross-cultural awareness of the ocean at all levels of public engagement and establish the crucial roles that oceans play in past, present, and future human well-being. This theme has important cultural and environmental impacts on both policy and behavior, and its success will depend on close coordination among many elements of society including industry, military, academic, health, philanthropic, and community-based sectors. The infrastructure and approaches established to monitor and understand urban seas in the United States could be used to model other systems around the world and form the basis for global collaborative projects. And, perhaps most importantly, a concerted effort to engage young learners from all backgrounds in "backyard" student science activities can bring the critical importance of maintaining ocean health to those who may not otherwise be aware of the wonders of the ocean.

Decade Challenges and Outcomes Addressed

Healthy Urban Seas could contribute to many (if not all) of the UN Ocean Decade challenges and outcomes. In particular, this theme addresses Challenge 2: "Protect and restore ecosystems and biodiversity - Understand the effects of multiple stressors on ocean ecosystems and develop solutions to monitor, protect, manage and restore ecosystems and their biodiversity under changing environmental, social and climate conditions"; Challenge 4: "Develop a sustainable and equitable ocean economy - Generate knowledge, support innovation and develop solutions for equitable and sustainable development of the ocean economy under changing environmental, social and climate conditions"; Challenge 5: "Unlock ocean-based solutions to climate change - Enhance understanding of the ocean-climate nexus and generate knowledge and solutions to mitigate, adapt and build resilience to the effects of climate change across all geographies and at all scales, and to improve services including predictions for the ocean, climate and weather"; Challenge 6: "Increase community resilience to ocean hazards - Enhance multi-hazard early warning services for all geophysical, ecological, biological, weather-, climate- and anthropogenic-related ocean and coastal hazards, and mainstream community preparedness and resilience"; Challenge 9: "Skills, knowledge and technology for all - Ensure comprehensive capacity development and equitable access to data, information, knowledge and technology across all aspects of ocean science and for all stakeholders"; and Challenge 10: "Change humanity's relationship with the ocean - Ensure that the multiple values and services of the ocean for human well-being, culture and sustainable development are widely understood, and identify and overcome barriers to behaviour change required for a step change in humanity's relationship with the ocean."[12] Most obvious is its relevance to Outcome 2: "A healthy and resilient ocean where marine ecosystems are understood, protected, restored and managed." However, the theme also addresses Outcome 4: "A predicted ocean where society understands and can respond to changing ocean conditions"; Outcome 5: "A safe ocean where life and livelihoods are protected from ocean-related hazards"; Outcome 6: "An accessible ocean with open and equitable access to data, information and technology and innovation"; and Outcome 7: "An inspiring and engaging ocean where society understands and values the ocean in relation to human well-being and sustainable development" (UNESCO-IOC, 2021a).

Connections to Ocean-Shots, UN Ocean Decade Actions, and U.S. Ocean Priorities

This theme encompasses ideas from more than 10 Ocean-Shots and overlaps with several UN Ocean Decade–endorsed projects, programs, and contributions (see Table 2.6). It also presents a "laboratory" for implementing the foundational themes of An Inclusive and Equitable Ocean and An Ocean of Data.

Potential Research Elements

- Development of an observing system that provides comprehensive coverage of key functional attributes. Elements could include fixed platforms in critical locations, mobile undersea platforms to surveil trouble spots, surface vessels or autonomous platforms, and fleets of specially outfitted drones.
- Improvement in the spatial and temporal monitoring of plastic waste leakage through the application of new technologies, such as remote sensing, autonomous underwater/remotely operated vehicles, sensor advances, passive samplers, and others (see NASEM, 2022c).
- Implementation and evaluation of citizen/community science initiatives, in combination with community-led activities, for data collection, based on individuals or civic infrastructure (e.g., water taxis, ferries, tunnels) as data-collecting actors.
- Development of better methodologies and design guidance for resilient urban infrastructure.
- Characterization and prediction of catastrophic failures of infrastructure particularly as it impacts the health of the ocean.

[12] See https://www.oceandecade.org/challenges.

TABLE 2.6 Connections of Healthy Urban Seas to Ocean-Shots and UN Ocean Decade Actions

Title
Ocean-Shots
EquiSea: The Ocean Science Fund for All
Feeding 10 Billion: Contributions from a Marine Circular Bioeconomy
Global Ocean and Human Health Program
TRITON: A Social Media Network for the Ocean
OceanCloud: Transforming Oceanography with a New Approach to Data and Computing
Marine Health Hubs: Building Interdisciplinary Regional Hubs of Excellence to Research and Address the Societal Impacts of Marine Debris Across the Globe
An Ocean Science Education Network for the Decade
Novel Coastal Ecosystems: Engineered Solutions to Accelerate Water Quality Restoration using Engineered Aeration
OceanPredict.US
An INFOstructure solution to the socio-ecological hazards of coastal flood control infrastructure
Revolutionizing Coastal Ocean Research through a Novel Share Model for the Long-term Sustainability of Humanity
UN Ocean Decade Endorsed Actions[a]
Fisheries Strategies for Changing Oceans and Resilient Ecosystems by 2030 (Decade Programme 63)
NSF Coastlines and People (Decade Contribution 135)
Estuarine Ecological Knowledge Network (EEKN) (Decade Project 43)
An Ocean Corps for Ocean Science (Decade Programme 9)
A multi-dimensional and inclusive approach for transformative capacity development (CAP-DEV 4 the Ocean) (Decade Project 39)

NOTES: See Appendix A, Table A.6, for a description of each. NSF, National Science Foundation.
[a] See https://www.oceandecade.org/decade-actions.

- Development and testing of models that encompass the physical, chemical, and biological characteristics of the system, including the concept of digital twins.
- Development of opportunities for outreach, inclusion, and co-development of knowledge with a particular focus on disadvantaged communities impacted by the degradation of coastal waters.

Studies on the response of the coastal ecosystem and key species to multiple stressors (e.g., pollutants in wastewater and run-off, underwater noise, nonnative species introductions, salinity and temperature fluctuations), and efficacy of remediation efforts.

- Acceleration of progress in adapting to new fuels and energy sources for use in ocean shipping.
- Equipping of ocean-going vessels coming in and out of our major seaports to collect data as part of a network of global ocean observation.

Potential Next Steps

A workshop to identify research opportunities that would benefit the health of urban seas and adjacent communities could be organized around topics or could focus on a specific urban port to highlight key issues such as the following:

- Opportunities and approaches for engaging coastal urban populations in the co-development of research priorities;
- Development and testing of new sensors and platforms (e.g., point pollution, repeat sampling drones)—in concert with The Ocean Revealed;
- Monitoring of pollutants, such as plastic and chemical contaminants, in coastal waters, and identification of inputs, fates, and effects;
- Improvement of predictive tools for hydrodynamic, sediment, and ecosystem modeling;
- Identification of educational opportunities to promote ocean culture and literacy;
- Demonstration projects for ecosystem-based resource management and entrepreneurship (e.g., seaweed harvesting, coastal blue carbon restoration [e.g., marsh, seagrass, mangrove habitats, wild capture fisheries, and aquaculture])—in coordination with The Restored and Sustainable Ocean; and
- Demonstration projects for enhancing ecosystems and reducing climate change risks through natural solutions and NBSs—in coordination with Ocean Solutions for Climate Resilience.

The ultimate objective of the workshop would be to describe the desired end state (i.e., what a healthy urban sea would look like) and then develop a bold and transformative program to achieve this end state.

Defining Success

By 2030, the urban sea will be well characterized through observation and modeling efforts leading to clean waters, enhanced coastal resilience, and greater community engagement. This will include the following:

- Infrastructure to support the blue economy.
- Publicly available quantification of the flux of important properties, such as pollutant concentrations, from one water body to another.

- Major stakeholders identified and engaged in monitoring, research, education, and management applications.
- An urban sea defined and understood through observational and modeling efforts, such as the establishment of a digital twin. For example, a well-sourced model could track and predict major environmental changes and their implications for socioeconomic activities. This could inform the design of infrastructure for coastal resilience.
- Measurable improvements in urban sea health through mitigation of pollution sources.
- Development of a community culture that understands the importance of, and plays an active role in, ensuring a safe and healthy ocean through community engagement (ocean cultural literacy).

3

Conclusions

Recognizing that the ocean is essential to the nation's economy, security, and well-being, the United States has been at the forefront of advancements in marine science and technology for the past century. The United Nations (UN) Decade of Ocean Science for Sustainable Development 2021–2030 (UN Ocean Decade) outcomes that call for a clean, healthy, resilient, productive, predicted, safe, accessible, and inspiring ocean in 2030 are consistent with the goals defined by many U.S. reports including *Science and Technology for America's Oceans: A Decadal Vision*, the federal plan which outlined five goals (understanding the ocean, promoting economic prosperity, ensuring maritime security, safeguarding human health, and developing resilient coastal communities) to advance ocean science and technology, and the nation, in the coming decade (see Table 1.1). Given the synergies between U.S. and UN Ocean Decade objectives, as well as the knowledge that the ocean connects and benefits all of humankind, the United States has a major interest in participating in what is perhaps a once-in-a-lifetime opportunity to engage the global community to provide scientific solutions for ocean sustainability.

In 2020, the National Oceanic and Atmospheric Administration asked the Ocean Studies Board at the National Academies of Sciences, Engineering, and Medicine (the National Academies) to establish the U.S. National Committee for the UN Decade of Ocean Science for Sustainable Development. Initially, the main role for the U.S. National Committee was to serve as the informational hub for the Ocean Decade–related activities taking place across the United States, a task accomplished with the establishment of a dedicated website, a regular newsletter, and public meetings and webinars. In addition, the U.S. National Committee issued a call for "Ocean-Shots"—transformative research concepts—as a way to engage and inspire the U.S. ocean science community to participate in the UN Ocean Decade.

In parallel, the UN Ocean Decade issued a call for programs and projects for endorsement by the international secretariat. In 2021, the United Nations endorsed 31 Programmes and 86 projects,[1] at least 30 of which were submitted by U.S. federal agencies and/or involved participation of U.S. scientists. Many of the endorsed programs involve efforts that are already established, at least partially funded, and ongoing, and as such provide a jump start for UN Ocean Decade activities. As outlined in the discussion of next steps below, relevant endorsed programs with U.S. participation will be included in all further development of the themes. In addition, the U.S. National Committee expects that some of the research developed under the themes will lead to new activities eligible for endorsement by the United Nations.

As the next step in advancing U.S. contributions to the UN Ocean Decade, the Subcommittee on Ocean Science and Technology requested a consensus study by the National Academies to identify between three and five cross-cutting themes that incorporate the most promising and innovative research concepts from the Ocean-Shot submissions. Many of the submitted Ocean-Shots addressed new approaches to the conduct of science to make it more inclusive and equitable, including providing greater access to ocean data and information. In response, this consensus committee developed two foundational themes—An Inclusive and Equitable Ocean and An Ocean of Data—and four topical themes, which represent areas where the committee identified opportunities to augment and advance ocean studies that support sustainable development as described in the UN Ocean Decade Implementation Plan. Some of the relationships between these initiatives are shown in Table 3.1.

The Ocean-Shots were not written to address the themes; rather, the themes were developed from the amalgamation and natural groupings of the Ocean-Shots. As a result, the identification of submissions that fit each theme was in many cases subjective and should not be interpreted as a comprehensive list of all of the Ocean-Shots that could contribute to a particular theme. Subsequent workshops addressing these themes will provide an opportunity for the Ocean-Shot authors and other members of the community to contribute their ideas to the development of research programs. The workshops will draw from a range of sectors including federal and state agencies, private industry and business, nongovernmental organizations, and foundations to help establish partnerships for theme development. In addition, these workshops will be organized to provide opportunities for interdisciplinary or multidisciplinary collaborations. The scope of each theme crosses disciplinary boundaries, requiring contributions from natural sciences, social sciences, and engineering as well as oceanography. The 2005 report *Facilitating Interdisciplinary Research* (NAS et al., 2005) recommends funding the collaborative process, not just the research team. In practice, this

[1] See https://www.oceandecade.org/decade-actions.

TABLE 3.1 UN Ocean Decade Outcomes and the Cross-Cutting Themes for U.S. Contributions to the Ocean Decade

UN Ocean Decade Outcomes (i.e., seven outcomes to describe the "ocean we want" at the end of the UN Ocean Decade)	Cross-Cutting Themes for U.S. Contributions to the Ocean Decade
A clean ocean	An Inclusive and Equitable Ocean
A healthy and resilient ocean	An Ocean of Data
A productive ocean	The Ocean Revealed
A predicted ocean	The Restored and Sustainable Ocean
A safe ocean	Ocean Solutions for Climate Resilience
An accessible ocean	Healthy Urban Seas
An inspiring and engaging ocean	

SOURCE: UNESCO-IOC, 2021a.

could include opportunities for researchers from various disciplines and sectors to meet for the purpose of enhancing communication and sharing knowledge and ideas, as envisioned here for the themed workshops. The goal is to foster scientific advances through the inclusion of the ideas and perspectives of fields beyond the traditional ocean sciences.

Several models have been developed to promote problem solving through interdisciplinary team research. The National Academies Keck Futures Initiative (NAKFI)[2] was designed to catalyze interdisciplinary research, using an approach that incorporated the findings from *Facilitating Interdisciplinary Research*. In this 15-year program, think-tank style conferences were organized around real-world challenges, with participants drawn from diverse backgrounds. The meetings were designed to provide venues for conversation, predominantly through the establishment of "seed idea groups" charged with developing solutions to the identified challenge. To extend the work beyond the conference, seed grants were awarded to participants to pursue ideas developed at the meeting analogous to start-up capital for a business. A second model is the incentive prize contest, such as that hosted by the XPRIZE foundation.[3] XPRIZE characterizes its contests as "competitions to crowdsource solutions to the world's grand challenges" (XPRIZE, n.d.). Innovators with diverse expertise are encouraged to form teams to solve the challenge and compete for a monetary prize. For example, the XPRIZE Carbon Removal challenges teams to find a method for carbon dioxide capture and sequestration "at a scale of at least 1000 tonnes removed per year; model their costs at a scale of 1 million tonnes per year; and show a pathway to achieving a scale of gigatonnes per year in future" (XPRIZE, n.d.). The reward for the winner is $100 million, funded by Elon Musk and the Musk Foundation. In both of these examples, the formation of interdisciplinary teams is motivated by the promise of funding, one at the start (NAKFI) and the other at the finish (XPRIZE).

[2] See http://www.nasonline.org/programs/keck-futures-initiative.
[3] See https://www.xprize.org.

Although the two foundational and four topical themes are presented separately, it is clear from even a cursory perusal that they are not independent and unquestionably overlap in several areas. The outcomes and best practices developed by the foundational themes are designed to be applied to each of the topical themes. However, the topical themes overlap and lessons can be learned that can be applied from one to the other. Many of the themes touch on the demand for information to fuel the blue (or ocean) economy. This includes increasing access to data and models as well as their analysis and interpretation. Application of the information to support sustainable development will require workforce training as well as increased ocean literacy, not just of the ocean environment and ecosystems but also the role of the ocean in the economy. The tools, techniques, and approaches developed under The Ocean Revealed will become critical components of each of the three other topical themes. Studies of increased coastal resilience under a variety of climate scenarios developed under Ocean Solutions for Climate Resilience are not independent of the coastal ecosystems examined under The Restored and Sustainable Ocean, and nature-based solutions are common to both. The instrumentation and observational components of The Ocean Revealed, the coastal processes and sea level modeling conducted under Ocean Solutions for Climate Resilience, and the ecosystem studies and modeling of The Restored and Sustainable Ocean are all components of Healthy Urban Seas. Additionally, several themes recognize the lack of information in relatively inaccessible regions including the Arctic Ocean, Southern Ocean, and deep ocean, yet the need to understand these regions becomes ever more critical as climate drives changes in physical, chemical, and biological processes. This is not unexpected—the components of the ocean are inextricably linked and cannot be compartmentalized (indeed ocean science has been impeded by this approach), and recognition of these connections up front will allow them to be used to advantage. For example, research groups from one theme could meet with another to address overlapping issues.

The committee has recommended two foundational and four topical themes that have the potential, if further developed and funded, to openly engage a broad, inclusive community and greatly advance our understanding of the ocean—its processes, resources, and values—to enable "the ocean we need for the future we want." The UN Ocean Decade represents a global opportunity to advance ocean science and sustainable practices; it is also an opportunity for the United States to demonstrate international leadership in furthering efforts to ensure that the ocean continues to support the well-being of communities at home and worldwide, for both current and future generations. Building on the efforts of the U.S. ocean community members who submitted Ocean-Shots, activities endorsed by the UN Ocean Decade process, and many complementary efforts such as OceanObs'19 and the High Level Panel reports, the committee offers these themes as the next step toward realizing the promise of the UN Ocean Decade.

References

AIAA (American Institute of Aeronautics and Astronautics). 2020. *Digital Twin: Definition and Value.* An AIAA and AIA Position Paper. https://www.aia-aerospace.org/report/digital-twin-paper.

Alder, J., B. Campbell, V. Karpouzi, K. Kaschner, and D. Pauly. 2008. Forage Fish: From Ecosystems to Markets. *Annual Review of Environment and Resources* (33):153–166.

Boukabara, S.-A., V. Krasnopolsky, S.G. Penny, J.Q. Stewart, A. McGovern, D. Hall, J.E.T. Hoeve, J. Hickey, H.-L.A. Huang, J.K. Williams, K. Ide, P. Tissot, S.E. Haupt, K.S. Casey, N. Oza, A.J. Geer, E.S. Maddy, and R.N. Hoffman. 2020. Outlook for Exploiting Artificial Intelligence in the Earth and Environmental Sciences. *Bulletin of the American Meteorological Society* 102(5):1–53. https://doi.org/10.1175/bams-d-20-0031.1.

Brundtland, G. 1987. *Report of the World Commission on Environment and Development: Our Common Future.* United Nations General Assembly Document A/42/427.

Buck, J.H., S.J. Bainbridge, E.F. Burger, A.C. Kraberg, M. Casari, K.S. Casey, L. Darroch, J. Del Rio, K. Metfies, E. Delroy, P.F. Fischer, T. Gardner, R. Heffernan, S. Jirka, A. Kokkinaki, M. Loebl, P.L. Buttigieg, J.S. Pearlman, and I. Schewe. 2019. Ocean Data Product Integration Through Innovation—the Next Level of Data Interoperability. Review. *Frontiers in Marine Science.* https://doi.org/10.3389/fmars.2019.00032.

Camps-Valls, G., D. Tuia, X.X. Zhu, and M. Reichstein, eds. 2021. *Deep Learning for the Earth Sciences: A Comprehensive Approach to Remote Sensing, Climate Science, and Geosciences.* Hoboken, NJ: John Wiley & Sons. https://doi.org/10.1002/9781119646181.

Drazen, J.C., C.R. Smith, K.M. Gjerde, S.H.D. Haddock, G.S. Carter, C.A. Choy, M.R. Clark, P. Dutrieux, E. Goetze, C. Hauton, M. Hatta, J.A. Koslow, A.B. Leitner, A. Pacini, J.N. Perelman, T. Peacock, T.T. Sutton, L. Watling, and H. Yamamoto. 2020. Opinion: Midwater Ecosystems Must Be Considered When Evaluating Environmental Risks of Deep-Sea Mining. *Proceedings of the National Academy of Sciences* 114(30):202011914–202017460.

EPA (U.S. Environmental Protection Agency). 2009. *Valuing the Protection of Ecological Systems and Services.* A Report of the EPA Science Advisory Board. Washington, DC: EPA.

FAO (Food and Agriculture Organization of the United Nations). 2020. *The State of World Fisheries and Aquaculture.* Rome, Italy: FAO. https://www.fao.org/3/ca9229en/online/ca9229en.html.

Gentemann, C.L., C. Holdgraf, R. Abernathey, D. Crichton, J. Colliander, E.J. Kearns, Y. Panda, and R.P. Signell. 2021. Science Storms the Cloud. *AGU Advances* 2(2):e2020AV000354-7. https://doi.org/10.1029/2020av000354.

GIDA (Global Indigenous Data Alliance) Research Data Alliance International Indigenous Data Sovereignty Interest Group. 2019. CARE Principles for Indigenous Data Governance. https://static1.squarespace.com/static/5d3799de845604000199cd24/t/5da9f4479ecab221ce848fb2/1571419335217/CARE+Principles_One+Pagers+FINAL_Oct_17_2019.pdf.

Gori, A., N. Lin, D. Xi, and K. Emanuel. 2022. Tropical Cyclone Climatology Change Greatly Exacerbates US Extreme Rainfall–Surge Hazard. *Nature Climate Change* 12:171–178. https://doi.org/10.1038/s41558-021-01272-7.

Hofstra, B., V.V. Kulkarni, S.M. Galvez, B. He, D. Jurafsky, and D.A. McFarland. 2020. The Diversity-Innovation Paradox in Science. *Proceedings of the National Academy of Sciences* 117(17):9284–9291.

IOC-UNESCO (Intergovernmental Oceanographic Commission-United Nations Educational, Scientific and Cultural Organization). 2020. *Non-Paper on Existing and Potential Future Services of the IOC-UNESCO in Support of a Future ILBI for the Conservation and Sustainable Use of Biodiversity Beyond National Jurisdiction (BBNJ).* Information document series 1387. Paris, France: IOC. https://unesdoc.unesco.org/ark:/48223/pf0000374421.locale=en.

IPCC (Intergovernmental Panel on Climate Change). 2018. Summary for Policymakers. In *Global Warming of 1.5°C. An IPCC Special Report on the Impacts of Global Warming of 1.5°C Above Pre-industrial Levels and Related Global Greenhouse Gas Emission Pathways, in the Context of Strengthening the Global Response to the Threat of Climate Change, Sustainable Development, and Efforts to Eradicate Poverty*, edited by V. Masson-Delmotte, P. Zhai, H.-O. Pörtner, D. Roberts, J. Skea, P.R. Shukla, A. Pirani, W. Moufouma-Okia, C. Péan, R. Pidcock, S. Connors, J.B.R. Matthews, Y. Chen, X. Zhou, M.I. Gomis, E. Lonnoy, T. Maycock, M. Tignor, and T. Waterfield. Geneva, Switzerland: World Meteorological Organization.

Kapteyn, N.G., and K.E. Willcox. 2021. Digital Twins: Where Data, Mathematics, Models, and Decisions Collide. *SIAM News*. https://sinews.siam.org/Details-Page/digital-twins-where-data-mathematics-models-and-decisions-collide.

Lueth, K.L. 2014. Why the Internet of Things Is Called the Internet of Things: Definition, History, Disambiguation. *IOT Analytics*. https://iot-analytics.com/internet-of-things-definition.

Mayer, L., M. Jakobsson, G. Allen, B. Dorschel, R. Falconer, V. Ferrini, G. Lamarche, H. Snaith, and P. Weatherall. 2018. The Nippon Foundation—GEBCO Seabed 2030 Project: The Quest to See the World's Oceans Completely Mapped by 2030. *Geosciences* 8:63. https://doi.org/10.3390/geosciences8020063.

MEA (Millennium Ecosystem Assessment). 2005. *Ecosystems and Human Well-Being: Current State and Trends. Coastal Systems.* Washington, DC: Island Press.

NAS, NAE, and IOM (National Academy of Sciences, National Academy of Engineering, and Institute of Medicine). 2005. *Facilitating Interdisciplinary Research.* Washington, DC: The National Academies Press. https://doi.org/10.17226/11153.

NASEM (National Academies of Sciences, Engineering, and Medicine). 2017. *Sustaining Ocean Observations to Understand Future Changes in Earth's Climate*. Washington, DC: The National Academies Press. https://doi.org/10.17226/24919.

NASEM. 2018a. *Understanding the Long-Term Evolution of the Coupled Natural-Human Coastal System: The Future of the U.S. Gulf Coast.* Washington, DC: The National Academies Press. https://doi.org/10.17226/25108.

NASEM. 2018b. *Open Science by Design: Realizing a Vision for 21st Century Research.* Washington, DC: The National Academies Press. https://doi.org/10.17226/25116.

NASEM. 2019. *A Research Review of Interventions to Increase the Persistence and Resilience of Coral Reefs*. Washington, DC: The National Academies Press. https://doi.org/10.17226/25279.

NASEM. 2020a. *Evolving the Geodetic Infrastructure to Meet New Scientific Needs*. Washington, DC: The National Academies Press. https://doi.org/10.17226/25579.

NASEM. 2020b. *Sustaining Ocean Observations: Proceedings of a Workshop—in Brief.* Washington, DC: The National Academies Press. https://doi.org/10.17226/25997.

NASEM. 2021. *Global Change Research Needs and Opportunities for 2022–2031.* Washington, DC: The National Academies Press. https://doi.org/10.17226/26055.

NASEM. 2022a. *A Research Strategy for Ocean-Based Carbon Dioxide Removal and Sequestration.* Washington, DC: The National Academies Press. https://doi.org/10.17226/26278.

NASEM. 2022b. *Next Generation Earth Systems Science at the National Science Foundation.* Washington, DC: The National Academies Press. https://doi.org/10.17226/26042.

NASEM. 2022c. *Reckoning with the U.S. Role in Global Ocean Plastic Waste.* Washington, DC: The National Academies Press. https://doi.org /10.17226/26132.

NASEM. 2022d. *Machine Learning and Artificial Intelligence to Advance Earth System Science: Opportunities and Challenges.* Washington, DC: The National Academies Press. https://doi.org/10.17226/26566.

NRC (National Research Council). 2007. *Mitigating Shore Erosion Along Sheltered Coasts.* Washington, DC: The National Academies Press. https://doi.org/10.17226/11764.

Ocean Panel. 2020. *Transformations for a Sustainable Ocean Economy: A Vision for Protection, Production and Prosperity.* https://www.oceanpanel.org/ocean-action/files/transformations-sustainable-ocean-economy-eng.pdf.

OceanObs'19. 2019. Conference Statement. https://www.oceanobs19.net/wp-content/uploads/2019/09/OO19-Conference-Statement_online.pdf.

Onink, V., C.E. Jongedijk, M.J. Hoffman, E. van Sebille, and C. Laufkötter. 2021. Global Simulations of Marine Plastic Transport Show Plastic Trapping in Coastal Zones. *Environmental Research Letters* 16(6). https://iopscience.iop.org/article/10.1088/1748-9326/abecbd.

Phillips, M. 2009. Mariculture Overview. In *Encyclopedia of Ocean Sciences, Second Edition*, edited by J.H. Steele, pp. 537–544. Oxford, UK: Academic Press.

Ramachandran, R., K. Bugbee, and K. Murphy. 2021. From Open Data to Open Science. *Earth and Space Science* 8(5):e2020EA001562. https://doi.org/10.1029/2020ea001562.

Reichstein, M., G. Camps-Valls, B. Stevens, M. Jung, J. Denzler, N. Carvalhais, and Prabhat. 2019. Deep Learning and Process Understanding for Data-Driven Earth System Science. *Nature* 566(7743):195–204. https://doi.org/10.1038/s41586-019-0912-1.

Samuel, R.M., R. Meyer, P.L. Buttigieg, N. Davies, N.W. Jeffery, C. Meyer, C. Pavloudi, K.J. Pitz, M. Sweetlove, S. Theroux, J. van de Kamp, and A. Watts. 2021. Toward a Global Public Repository of Community Protocols to Encourage Best Practices in Biomolecular Ocean Observing and Research. Perspective. *Frontiers in Marine Science* 8:758694. https://doi.org/10.3389/fmars.2021.758694.

SOST (Subcommittee on Ocean Science and Technology of the National Science and Technology Council). 2018. *Science and Technology for America's Oceans: A Decadal Vision.* Washington, DC. https://oeab.noaa.gov/wp-content/uploads/2020/Documents/Science-and-Technology-for-Americas-Oceans-A-Decadal-Vision.pdf.

Stuchtey, M., A. Vincent, A. Merkl, M. Bucher, P.M. Haugan, J. Lubchenco, and M.E. Pangestu. 2020. Ocean Solutions That Benefit People, Nature and the Economy. Washington, DC: World Resources Institute. www.oceanpanel.org/ocean-solutions.

Tacon, A., and M. Metian. 2008. Global Overview on the Use of Fish Meal and Fish Oil in Industrially Compounded Aquafeeds: Trends and Future Prospects. *Aquaculture* 285(1–4):146–158.

Thatcher, C.A., J.C. Brock, and E.A. Pendleton. 2013. Economic Vulnerability to Sea-Level Rise Along the Northern U.S. Gulf Coast. *Journal of Coastal Research* 63:234–243. https://doi.org/10.2112/si63-017.1.

UNESCO (United Nations Educational, Scientific and Cultural Organization). n.d. Facts and Figures on Marine Pollution. http://www.unesco.org/new/en/natural-sciences/ioc-oceans/focus-areas/rio-20-ocean/blueprint-for-the-future-we-want/marine-pollution/facts-and-figures-on-marine-pollution.

UNESCO-IOC (Intergovernmental Oceanographic Commission). 2021a. *The United Nations Decade of Ocean Science for Sustainable Development (2021–2030) Implementation Plan*. IOC Ocean Decade Series, 20. Paris, France: UNESCO. https://www.oceandecade.org/wp-content/uploads//2021/09/337567-Ocean%20Decade%20Implementation%20Plan%20-%20Full%20Document.

UNESCO-IOC. 2021b. *National Decade Committees Operational Guidelines*. IOC Ocean Decade Series, 24. Paris, France: UNESCO. https://www.oceandecade.org/wp-content/uploads//2021/09/337557-National%20Decade%20Committees%20Operational%20Guidelines.

USGS (U.S. Geological Survey). 2018. Environmental DNA (eDNA). https://www.usgs.gov/special-topics/water-science-school/science/environmental-dna-edna.

Waterston, J. n.d. Ocean of Things. Defense Advanced Research Projects Agency. https://www.darpa.mil/program/ocean-of-things.

Weller, R.A., D.J. Baker, M.M. Glackin, S.J. Roberts, R.W. Schmitt, E.S. Twigg, and D.J. Vimont. 2019. The Challenge of Sustaining Ocean Observations. *Frontiers in Marine Science* 6:717–718. https://doi.org/10.3389/fmars.2019.00105.

Wilkinson, M.D., M. Dumontier, I.J. Aalbersberg, G. Appleton, M. Axton, A. Baak, N. Blomberg, J.-W. Boiten, L. Bonino da Silva Santos, P.E. Bourne, J. Bouwman, A.J. Brookes, T. Clark, M. Crosas, I. Dillo, O. Dumon, S. Edmunds, C.T. Evelo, R. Finkers, A. Gonzalez-Beltran, A.J.G. Gray, P. Groth, C. Goble, J.S. Grethe, J. Heringa, P.A.C 't Hoen, R. Hooft, T. Kuhn, R. Kok, J. Kok, S.J. Lusher, M.E. Martone, A. Mons, A.L. Packer, B. Persson, P. Rocca-Serra, M. Roos, R. van Schaik, S.-A. Sansone, E. Schultes, T. Sengstag, T. Slater, G. Strawn, M.A. Swertz, M. Thompson, J. van der Lei, E. van Mulligen, J. Velterop, A. Waagmeester, P. Wittenburg, K. Wolstencroft, J. Zhao, and B. Mons. 2016. The FAIR Guiding Principles for Scientific Data Management and Stewardship. *Scientific Data* 3:160018. https://www.nature.com/articles/sdata201618.

World Bank and United Nations Department of Economic and Social Affairs. 2017. *The Potential of the Blue Economy: Increasing Long-Term Benefits of the Sustainable Use of Marine Resources for Small Island Developing States and Coastal Least Developed Countries*. Washington, DC: World Bank.

World Commission on Environment and Development. 1987. *Our Common Future*. Oxford, UK: Oxford University Press.

XPRIZE. n.d. $100M Prize for Carbon Removal. https://www.xprize.org/prizes/elonmusk.

Acronyms and Abbreviations

AI	artificial intelligence
AMOC	Atlantic Meridional Overturning Circulation
CARE	collective benefit, authority to control, responsibility, and ethics
CDR	carbon dioxide removal
eDNA	environmental deoxyribonucleic acid
EU	European Union
FAIR	findable, accessible, interoperable, and reusable
IOC	Intergovernmental Oceanographic Commission
LDC	Least Developed Country
ML	machine learning
NAKFI	National Academies Keck Futures Initiative
NBS	nature-based solution
NSTC	National Science and Technology Council
SIDS	Small Island Developing States
SOST	Subcommittee on Ocean Science and Technology
UN	United Nations
UNESCO	United Nations Educational, Scientific and Cultural Organization

Glossary

Atlantic Meridional Overturning Circulation: large system of ocean currents in the Atlantic Ocean that transport warm, salty water from the tropics northward.

Blue/ocean economy: defined by the World Bank and the United Nations Department of Economic and Social Affairs (2017) as "sustainable use of ocean resources for economic growth, improved livelihoods and jobs, and ocean ecosystem health." Also, the World Bank and the United Nations Department of Economic and Social Affairs (2017) describes the blue economy as a concept: "to promote economic growth, social inclusion, and the preservation or improvement of livelihoods while at the same time ensuring environmental sustainability of the oceans and coastal areas."

Citizen/community science: public participation, in part or in full, in scientific research.

Coastal blue carbon: nature-based sequestration of carbon dioxide in coastal ecosystems such as mangroves, salt marshes, and seagrasses.

Coastal communities: the neighborhoods, towns, and cities located on shorelines and adjacent lands.

Digital twin: a "set of virtual information constructs that mimics the structure, context and behavior of an individual/unique physical asset, or a group of physical assets, is dynamically updated with data from its physical twin throughout its life cycle and informs decisions that realize value" (AIAA, 2020). A digital twin is highly dynamical, mimicking the time evolution of its physical asset (PA) via advanced simulation and emulation capabilities; it is updated by ingesting vast amounts of observational data of diverse types; and it enables WHAT-IF queries and multiple realizations to support prediction of responses of the PA to hypothetical perturbations with quantified uncertainties (Kapteyn and Willcox, 2021).

Ecosystem services: defined by the U.S. Environmental Protection Agency as "the direct or indirect contributions that ecosystems make to the well-being of human populations" (EPA, 2009) and defined by the Millennium Ecosystem Assessment as "the benefits people obtain from ecosystems" (MEA, 2005).

Environmental DNA (eDNA): defined by the U.S. Geological Survey as "organismal DNA that can be found in the environment." "Environmental DNA originates from cellular material shed by organisms (via skin, excrement, etc.) into aquatic or terrestrial environments that can be sampled and monitored using new molecular methods" (USGS, 2018).

Hydrokinetic energy conversion: the transformation of kinetic energy from inland rivers, tidal estuaries and channels, and ocean currents or waves into electricity.

Interdisciplinary research: "a mode of research by teams or individuals that integrates information, data, techniques, tools, perspectives, concepts, and/or theories from two or more disciplines or bodies of specialized knowledge to advance fundamental understanding or to solve problems whose solutions are beyond the scope of a single discipline or area of research practice" (NAS et al., 2005, p. 2).

Mariculture: the *Encyclopedia of Ocean Sciences, Second Edition*, defines mariculture as "the farming of marine organisms for food and other products such as pharmaceuticals, food additives, jewelry (e.g., cultured pearls), nutraceuticals, and cosmetics, either in the natural marine environment, or in land- or sea-based enclosures, such as cages, ponds, or raceways" (Phillips, 2009).

Ocean state estimation: formal approaches using methods from statistical estimation or control theory to extract information from (generally incomplete) observations, (generally uncertain) numerical models, and theory, to derive complete, optimal descriptions of the time-evolving state of the ocean and its uncertainty.

Open access data: publicly available data that can be found online free of charge; free to download, analyze, process, or modify; and free to reuse or redistribute without restrictions. Open access data often come with public open data licenses;

common examples are the Creative Commons Attribution (CC-BY) license or the Open Database License (ODbL).

Relocation: the intentional and coordinated movement of people, assets, and/or infrastructure away from threats of sea level rise, climate-driven floods, and more; for the purpose of this report, relocation is synonymous with managed retreat.

Sustainable development: as defined in the Brundtland Commission report, "development that meets the needs of the present without compromising the ability of future generations to meet their own needs." It encompasses the pathways and processes necessary to achieve sustainability (World Commission on Environment and Development, 1987).

Appendix A

Connections to Ocean-Shots and United Nations Ocean Decade Actions: Tables

The committee referenced the Ocean-Shots and the United Nations (UN) Ocean Decade–endorsed actions in the formulation of each of the foundational and topical themes. For the UN Ocean Decade–endorsed actions, the committee focused on those that have prominent U.S. participation as a first step toward integrating new U.S. initiatives, identified through the Ocean-Shots, with the international efforts endorsed by the United Nations. However, the committee recognizes that other UN-endorsed actions offer opportunities for collaboration with U.S. researchers and encourage project teams to review the full list of UN-endorsed actions to identify potential collaborators in the international community.

The Ocean-Shot Decade Directory[1] is populated with the "Titles" and "Authors" as submitted by the authors. The descriptions are either sourced directly from the summaries of the Ocean-Shots or paraphrased by National Academies of Sciences, Engineering, and Medicine staff based on the abstracts or content of the Ocean-Shots.

The tables below contain the titles and descriptions (i.e., summaries) for a subset of Ocean-Shots and UN Ocean Decade–endorsed actions relevant to each foundational or topical theme. The titles and summaries of the latter were sourced directly from the UN Ocean Decade's *Endorsed Decade Actions.*[2]

---

[1] See https://www.nationalacademies.org/our-work/us-national-committee-on-ocean-science-for-sustainable-development-2021-2030/ocean-shot-directory#:~:text=%E2%80%9COcean%2DShots%E2%80%9D%20are%20ambitious,ocean%20science%20for%20sustainable%20development.

[2] An Excel file of the PDF version of the document can be downloaded from https://oceanexpert.org/document/29188.

TABLE A.1 Connections of the Foundational Theme An Inclusive and Equitable Ocean to Ocean-Shots and UN Ocean Decade Actions

Title	Description
Ocean-Shots	
FantaSEAS Project: Incorporating Inspiring Ocean Science in the Popular Media	Promoting ocean literacy and diversity in ocean science, by engaging industries and the general public through art, television, and literature.
EquiSea: The Ocean Science Fund for All	EquiSea aims to improve equity in ocean science by establishing a philanthropic fund to provide direct financial support to projects, coordinating capacity development activities, fostering collaboration and co-financing of ocean science between academia, government, NGOs, and private sector actors, and supporting the development of low-cost and easy-to-maintain ocean science technologies.
The Ocean Decade Show	The Ocean Decade Show is your monthly source of behind-the-scenes information about the United Nations Decade of Ocean Science for Sustainable Development. This podcast will be a guide exploring the history, planning, preparation, and execution of the Ocean Decade.
The Estuarine Ecological Knowledge Network (EEKN): The View from SE Louisiana and Future Prospects	Increase communication between coastal community members, scientists, and policymakers in Pilot Project Sites.
TRITON: A Social Media Network for the Ocean	A different, unified approach to outreach and communication: TRITON, a social media platform. The platform will serve as a one-stop-shop for information about the ocean and society, a place where scientists and organizations can share and amplify science-based information and for dissemination of original content that connects people to the ocean through the lens of their own community experiences.
Ocean Technology Field Academy	The Ocean Technology Field Academy will support ocean-focused communities and advance ocean understanding by providing competency-driven micro-credentials representing a participant's ability to exploit, enhance and promote sensors, sensor platforms, sensor networks, crewed and uncrewed underwater vehicles, sonar systems, and process data.
Small Islands, Big Impact	Global declines in ocean health combined with the upheaval of COVID-19 are unraveling Pacific Island states' social, economic, and cultural fabric. This concept is designed to co-develop science innovation and ocean solutions that foster crisis readiness and are scalable to multiple geographies.
ICOFS (Integrated Coastal Ocean Forecast Systems)	ICOFS: Integrated Coastal Ocean Forecast Systems is the U.S. component of the international "CoastPredict: Observing and Predicting the Global Coastal Ocean" Programme coastpredict.org that has been proposed to the Decade and is being developed as a component of the IOC/GOOS (Global Ocean Observing System).

TABLE A.1 Continued

Title	Description
An Ocean Science Education Network for the Decade	A coordinated ocean science education network will leverage resources across sectors and nations to reach global citizenry. A flow of information among scientists, education professionals, indigenous leaders, policymakers, business leaders, and the public will help guide research priorities and enhance global ocean literacy.
Building Ocean Collaborations	The project's objectives are to sustain ocean observations, identify a funding model for special projects, support pan-regional products, super-regional services, connect ocean programs across sectors, and meet Administration priorities.
Envisioning an Interconnected Ocean: Understanding the Links Between Geological Ocean Structure and Coastal Communities in the Pacific	The Ocean Explorations Trust (OET) will conduct scientific expeditions to better understand the ocean through seafloor mapping and ocean exploration. OET seeks to collaborate with local communities to reveal the structural significance and interconnected nature of oceanic features, making a link to the livelihoods of Pacific islanders.
Ocean Memory Project: A Cross-Disciplinary Approach to Global Scale Challenges	Ocean Memory Project's (OMP) vision is to grow a network of interconnected regional or themed nodes, each able to engage with an array of local challenges and partners while remaining in dialog with our larger community of thought and practice, thus allowing for engaged community growth that links people, and a growing knowledge base, from local to a global scale.
The 4Site Pacific Transect Collaborative	Combining scientific and culturally-grounded data, knowledge, values, and perspectives to equitably achieve Sustainable Development Goals and resilient wellbeing of coastal social-ecological systems.
Just, Equitable, Diverse, and Inclusive (JEDI) Aquanautics: Democratizing Innovation in the Networked Blue Economy	JEDI Aquanautics will fuse the world's most powerful experiential supercomputer, the NSF Holodeck, with Ocean Space Habitat's SM transformative capabilities—creating a world-class sociotechnical convergence catalyst for the Networked Blue Economy (NBE). JEDI "Inventioneering" will foster open-access passion-based convergence research, education, and innovation to inspire and train the next generation of lifelong learners and innovators. The overarching project vision is to lay the foundations of a vibrant JEDI Aquanautics ecosystem across academia, industry, government, public and private organizations, diverse multi-stakeholder communities and end-users. Significantly increasing human engagement with ocean environments, JEDI Aquanautics will deliver profound benefits throughout the global NBE.

continued

TABLE A.1 Continued

Title	Description
Revolutionizing Coastal Ocean Research through a Novel Share Model for the Long-term Sustainability of Humanity	We propose a bold vision for conducting ocean science. While the coastal ocean is essential to the well-being and long-term sustainability of humanity, our scientific knowledge on coastal ecosystems falls woefully short. We are calling for a new model that addresses social, economic, political, and logistical factors that impede participation in coastal ocean science through instruments and community relationships. Our new model will build on familiar tenets of the share economy developed to respond to the needs of coastal ocean research and science.
An Ocean Corps for Ocean Science (also a UN Ocean Decade Endorsed Programme)	Forming collaborations between US scientists and scientists in under-resourced countries can help address the unequal distribution of ocean science infrastructure worldwide. Just as the Peace Corps inspired young Americans, Ocean Corps will inspire US oceanographers to engage the world.
UN Ocean Decade Endorsed Actions[a]	
Ocean Voices: Building transformative pathways to achieve the Decade's outcomes (Decade Programme 16)	The contribution of ocean science to sustainable development is determined by people. Understanding the actors involved, their culture and wellbeing, and how power dynamics and decision-making processes influence our oceans is crucial to achieve the goals of the Decade and ensure knowledge, strategy and governance frameworks enable all to participate in, contribute to and benefit equitably from the Decade. The Oceans for Everyone program will conduct research, incubate ideas, facilitate critical discussions and convene capacity building partnerships to identify barriers and pathways and enabling conditions for equity in the Decade.
An Ocean Corps for Ocean Science (Decade Programme 9)	Motivated by the example of the US Peace Corps, we propose "An Ocean Corps for Ocean Science" as a unifying concept for sustaining long-term education and research collaborations between scientists from under-resourced nations and higher-resourced nations. Based upon our experience in Peace Corps and with the Coastal Ocean Environment Summer School in Ghana (https://coessing.org), we are confident that an Ocean Corps would inspire large numbers of scientists, especially early-career scientists, into its ranks, thus "internationalizing" their outlook, molding many of them into champions for international capacity development for the remainder of their careers, and fostering true ocean science collaborations worldwide.

TABLE A.1 Continued

Title	Description
AGU's Mentoring365: UN Decade of Ocean Sciences (Decade Contribution 226)	AGU's Mentoring365: UN Decade of Ocean Sciences increases equitable access to mentoring through a virtual global peer-to-peer mentorship program that will support and cultivate the career pipeline of those engaging in the Decade, particularly through the Early Career Ocean Professionals Programme. M365 matches and provides mentors and mentees with structured, relationship-building tools to develop and accomplish focused career goals. These connections range from one-on-one to small group discussions and often transcend national borders to bring a global perspective to the mentoring experience. The mentoring platform can be integrated into the online experience for specific Decade events or related activities including Ocean Sciences Meeting, Ocean Visions Summit, and AGU's Annual Fall Meeting with more than 25,000 scientists in attendance. In addition, the platform can also leverage other training capacities of AGU to further enhance the mentoring experience, such as professional trainings on communicating effectively with policy makers, the media, and public as well as fellowship opportunities through the Thriving Earth Exchange to learn how to work directly with local communities to co-create solutions. In discussions with representatives of the Early Career Ocean Professionals Programme, there was strong interest in leveraging M365 to support their programming needs and they have provided a letter of support for this proposal.
A Multi-Dimensional and Inclusive Approach for Transformative Capacity Development (CAP-DEV 4 the Ocean) (Decade Project 39)	This initiative brings together a diverse multi-national partnership to address the critical need for capacity development in ocean science, literacy and conservation in low- and middle-income countries. We seek to strengthen systems (funding, mentorship, recognition, learning, communication and evaluation) that address the challenges of the Ocean Decade and support the capacity to deliver ocean science and leadership, which is central to healthy oceans and sustainable development. This will proactively advance the integration of capacity development in the design and implementation of priorities for the Ocean Decade.

NOTE: AGU, American Geophysical Union; IOC, Intergovernmental Oceanographic Commission; NGO, nongovernmental organization; NSF, National Science Foundation.

[a] See https://www.oceandecade.org/decade-actions.

TABLE A.2 Connections of the Foundational Theme An Ocean of Data to Ocean-Shots and UN Ocean Decade Actions

Title	Description
Ocean-Shots	
Advancing Ocean Science through Open Science and Software on the Cloud	Seek to advance community awareness and practice around open science by taking advantage of innovative development tools and environments that will better enable science by decreasing barriers to collaborations, reproducibility, interdisciplinary research. For data, this would mark a shift from the central repository model to the central service model, enabling data-proximate computing.
Boundary Ocean Observation Network for the Global South (BOON-GS)	An existing network of established time series transects and areas collecting long-term data sets. The UN should build on the success of BOON and establish a BOON for the Global South. Increased observation coverage will improve understanding of the Ocean-Climate System.
COVERAGE: Next Generation Data Service Infrastructure for a Digitally Integrated Ocean Observing System in Support of Marine Science and Ecosystem Management	Coordinated international, multi-agency effort seeking to implement the next generation value-added data service infrastructure necessary to power a digitally integrated ocean observing system.
FathomNet: Exploring Our Ocean Using Artificial Intelligence	The adoption of AI in the ocean is limited by the availability of curated data, particularly underwater imagery and video, to train algorithms. FathomNET seeks to automate the processing of underwater imagery and video, to fully explore and discover our ocean.
Forward-Looking Decision Making in Fisheries in the Face of Climate Change	Bringing together a transdisciplinary team to transform data, capacity for predictive modeling, and the effectiveness of fisheries decision making. This team will work closely with stakeholders to design products that address key questions and help them make forward-looking climate-informed decisions.
Improved Value of the Observing System through Integrated Satellite and in situ Design	Over the decade this work will demonstrate the value of sustained remotely sensed data to the global ocean observing system and how uncertainties can be overcome through better-informed system design.
OceanCloud: Transforming Oceanography with a New Approach to Data and Computing	OceanCloud is a new vision for data infrastructure with three pillars: 1. Analysis-ready data shared openly in the cloud 2. On-demand, scalable distributed processing 3. Interactive, data-proximate computing, available to anyone

TABLE A.2 Continued

Title	Description
OceanPredict.US	Leveraging national and international collaborations and partnerships, OceanPredict.US aims to coalesce national global and basin-scale ocean assessment and prediction capabilities into operational implementations supporting nested regional and coastal activities. OceanPredict.US targets the full operational oceanography value chain, spanning: needs, observations, data management, analysis, ocean prediction, dissemination via a digital ocean, and information service delivery, culminating in end use while enabling innovation and value-added development.
An INFOstructure solution to the socio-ecological hazards of coastal flood control infrastructure	Our Ocean Shot concept is towards creating an information structure (Info-structure) to support decision making surrounding the use of engineering infrastructure for flood hazards in coastal areas. Often, use of such infrastructure occurs in the absence of socio-ecological data about coastal ecosystems and communities and result in unforeseen and multifaceted coastal disasters. The concept will be framed around case studies in the U.S. Gulf South and The Netherlands that highlight the increasing and changing use of flood control structures in response to climate impacts. The concept is scalable other ecosystems and geographies facing climate change related flooding.
UN Ocean Decade Endorsed Actions[a]	
Deep Ocean Observing Strategy (Decade Programme 129)	DOOS represents an interconnected network of deep-ocean observing, mapping, exploration, and modelling programs working together for the coming decade to 1) characterize the physics, biogeochemistry and biology of the deep ocean in space and time, 2) establish a baseline required to understand changes to its habitats and services, and 3) provide the information needed to have a healthy, predicted, resilient and sustainably-managed (deep) ocean. DOOS will promote the human capital and observing infrastructure needed to address critical scientific and management questions related to the climate, biodiversity and sustainability, while growing a diverse and inclusive next generation of deep-ocean leaders.

continued

TABLE A.2 Continued

Title	Description
The World Ocean Database Programme (WODP) (Decade Contribution 122)	Countries need access to oceanographic profile data of known quality to address current and emergent scientific and socio-economic issues at all spatial and temporal scales. The challenge is that data users cannot access the immense and growing globally distributed data that exists in diverse digital formats. The World Ocean Database (WOD) mitigates this challenge. WOD is the world largest unrestricted, uniformly formatted, quality controlled, digital ocean profile database available with data from 1778 to present. WOD acquires and receives ocean data worldwide for ocean Climate Essential Variables (EOV), plankton, and other variables including data from the World Data Service for Oceanography; part of the World Data System. WOD is hosted at NOAA and it is a project of the International Oceanographic Data Exchange (IODE) of the Intergovernmental Oceanographic Committee (IOC). WOD is a Center for Marine Meteorology and Oceanographic Climate Data (CMOC) in the Marine Climate Data System; a joint system of IOC and the World Meteorological Organization (WMO). In partnership with IODE, NOAA proposes to develop and deploy a data ingestion tool at the IODE project office in Belgium together with Cloud services as a starting point. This effort builds on IODE's Ocean Data and Information System (ODIS) and will enable National Oceanographic Data Centers worldwide and other digital repositories to (i) upload their oceanographic data into WOD and (ii) retrieve data in an uniform interoperable format; a value-added proposition. The vision is to achieve openly discoverable, accessible, and adaptable digital profile oceanographic data of known quality.
WOC SMART Ocean-SMART Industries (SO-SI): Science/ Industry Partnerships for Data Collection and Sharing (Decade Project 83)	The WOC SMART Ocean-SMART Industries (SO-SI) Program is the global, multi-industry initiative creating partnerships between science and the ocean private sector on data collecting and sharing. The SO-SI Program for the UN Ocean Decade will improve and scale this by developing the 1) Community of Practice - Engaging tens of thousands of ocean companies around data collecting and sharing 2) Clearinghouse and Knowledge Base - Assembling information from science on data collection programs, priorities, technologies and from industry on opportunities for using vessels and platforms 3) Matchmaking/Brokering Capacity - Expanding the efforts to connect science priorities with industry potential.

TABLE A.2 Continued

Title	Description
OneArgo: an integrated global, full depth and multidisciplinary ocean observing array for beyond 2020 (Decade Project 114)	OneArgo will transform the revolutionary "core" Argo array (which tracks the upper ocean physical state) to one that has truly global reach, including the polar oceans and marginal seas, extending to the full ocean depth and including ocean biogeochemical measurements. Through Argo's novel data management system, all data will be freely shared in real-time with a very high quality version delivered within 12 months. Implementing OneArgo will greatly increase Argo's already remarkable impact on ocean and climate services, predictions and research, and enable groundbreaking developments in understanding ocean ecosystems, forecasting ocean productivity and constraining the global carbon and energy budgets.
GO-SHIP Evolve (Decade Project 3)	"GO-SHIP Evolve" will focus on emerging opportunities for the fourth decadal GO-SHIP survey in: 1. New science areas: - Addition of biological and ecosystem measurements which will enable GO-SHIP to determine trends and variability in key indicators of ocean health. 2. Accelerating use of knowledge: - Expanding provision of real-time data. - Fit-for purpose provision of calibration data with evolving partner networks. - Creating new stakeholder-driven data products. 3. Multi-national capacity enabling contributions from more countries: - Developing access and training opportunities. - Supporting and facilitating multinational voyages and shared facilities. - Exploring new funding models for program-wide support and capacity.
Digital Twins of the Ocean - DITTO (Decade Programme 137)	DITTO will establish and advance a digital framework on which all marine data, modelling and simulation along with AI algorithms and specialized tools including best practice will enable shared capacity to access, manipulate, analyse and visualise marine information. It will enable users and partners to create ocean related development scenarios addressing issues such as energy, mining, fisheries, tourism and nature-based solutions. Digital-Twins can quantify benefits and environmental change and provide powerful visualizations. DITTO will empower ocean professionals including scientific users to create their own local or topical digital twins of "their ocean issue" by using standard workflows.

NOTE: NOAA, National Oceanic and Atmospheric Administration; WOC, World Ocean Council.
[a] See https://www.oceandecade.org/decade-actions.

TABLE A.3 Connections of the Topical Theme The Ocean Revealed to Ocean-Shots and UN Ocean Decade Actions

Title	Description
Ocean-Shots	
Long-Term, Global Seafloor Seismic, Acoustic and Geodetic Network	Establishment and maintenance of autonomous ships to collect and send data, including seafloor mapping, ocean sampling, and seafloor system maintenance ashore.
Unlocking the secrets of the evolving Central Arctic Ocean Ecosystem: A foundation for successful conservation and management	The Central Arctic Ocean is changing as fast as any other region, but understanding of the ecosystem is inadequate for effective protection, conservation, and management. A comprehensive expeditionary and autonomous approach will help quantify uncertain biological characteristics and rates over pan-Arctic spatial and temporal scales. This need is particularly timely given the recent international ratification of the "International Agreement to Prevent Unregulated Fishing in the High Seas of the Central Arctic Ocean," which has now come in to force. Meeting the challenges of the Agreement will require multiple nations and diverse stakeholders, including Indigenous coastal communities, to work together.
Arctic Shelves: Critical Environments in Flux	Arctic Seafloor Observatory Program: A new paradigm for ocean observing in a critical region. This concept supports the development of new approaches to sustain long-term seafloor observatories suitable for pan-Arctic shelf and ecosystem research.
Integrated Ocean Observing Across the Northwest Atlantic	We envision an interconnected system to track oceanographic and ecological changes from the Arctic to the Gulf of Maine, integrating state-of-the-art technologies, Indigenous knowledge, and citizen science to better understand the rapidly changing Northwest Atlantic.
Ocean Arc: An Ocean Shot for the Arctic	Photography can be a powerful instrument for change. Combining scientific research, new imaging technologies, Artificial Intelligence, and underwater photography, we'll share near to real-time changes of biodiversity in the Arctic and highlight these rapidly changing and unique environments.
Challenger150: A Global Survey of Deep Sea Ecosystems to Inform Sustainable Management	A blueprint for an inclusive, global, Deep-Sea field program. Emerged from Deep Ocean Stewardship working groups, the project's goal is to coordinate deep ocean research around the globe. With coordination, the project will harness the ability to use ocean knowledge to get us to the Deep Ocean that is wanted.

TABLE A.3 Continued

Title	Description
DORIS: Deep Ocean Research International Station	DORIS will map the Ocean's dynamic currents and patterns of marine life migration in a systematic manner using multispectral sensors deployed aboard the station, and with extra-vehicular arrays of edge-computing devices, Distributed Acoustic Sensing (DAS) arrays, and Uncrewed Underwater Vehicles (UUV).
Observing the Oceans Acoustically	Our vision is a global-spanning multi-purpose ocean acoustic network in direct analogy with GPS required to transform use and observation of the world below the ocean surface. A small number of judiciously-placed low-frequency acoustic sources transmitting to globally distributed receivers enable monitoring of acoustic geo-positioning, ocean measurements, and ocean sound time series.
Ocean Sound Atlas	The Ocean Sound Atlas will be a digital global ocean sound map. This interactive system will compile and integrate passive acoustic data by location of recording, for use by researchers, educators, policy-makers, engineers, explorers, sound artists, and other stakeholder groups.
Auscultating the Oceans: Developing a Marine Stethoscope	Auscultate is to examine a patient by listening to sounds. Just like doctors need stethoscopes, we need tools to measure ocean health.
Implementing a Global Deep Ocean Observing Strategy (iDOOS)	The Deep Ocean Observing Strategy (DOOS) is a GOOS project envisioning a globally integrated network of systems that observes the deep ocean (> 200m, with emphasis > 2000m) in support of strong science, policy and planning for sustainable oceans. This ocean shot (iDOOS) will implement an interconnected network of deep-ocean observing, mapping, exploration, and modelling programs working together for the coming decade.
SMART Subsea Cables for Observing the Ocean and Earth	SMART Cables' vision is to observe the oceans and Earth with a planetary-scale network of sensor-enabled submarine telecommunications cables to support climate, ocean circulation, sea-level monitoring, and tsunami and earthquake early warning and disaster risk reduction.
Measuring the Pulse of Earth's Global Ocean	We plan to deploy a unique, deep-ocean capable hydrophone-lander system at each of these deep-sea sites (>7k meters) Our goal is to make the first, simultaneous measurement of baseline ocean sound levels in what should be the quietest (i.e., lacking human-made noise) locations on Earth.

continued

TABLE A.3 Continued

Title	Description
The Endless Dive: Marine Species 3D response to climate change in oceans	It is well known throughout different ecological systems that with a warming climate, plant and animal species have moved northward to combat the changing climate. This has also been seen with respect to elevation, from reptilian microhabitats to large-scale mountain ranges. Could marine life follow this pattern? With such a unique, 3D landscape, species can move both deeper and northward in response to climate change. This could lead to impacts on fisheries as well as food chains and communities.
Accelerating Global Ocean Observing: Monitoring Coastal Ocean Through Broadly Accessible, Low-Cost Sensor Networks	This proposal seeks to develop a global network of low-cost, easily produced, and readily deployed ocean graphic censors for use on a wide variety of platforms in the coastal ocean.
Great Global Fish Count by DNA	Information on species diversity & abundance from DNA in 1 liter of water is comparable to information from 66 million liters trawled by a net. Time for the Great Global Fish Count! A project to collect data on millions of species through sediments in water.
A Global eDNA Monitoring System (GeMS)	GeMS would harness the readings provided by Environmental DNA, genetic material suspended in seawater. Molecular techniques can be used to see who was there, from viruses to top predators. The technology provides many opportunities to further understand ocean life and how to protect its species.
The US Ocean Biocode	The Ocean Biocode seeks to create a comprehensive DNA sequence library for all marine species by 2030. Marine biodiversity is the heart of ecosystems. Advancing DNA sequencing abilities give us the chance to sequence more, and now we need to translate this gathered information to species.
Boundary Ocean Observation Network for the Global South (BOON-GS)	An existing network of established time series transects and areas collecting long-term data sets. The UN should build on the success of BOON and establish a BOON for the Global South. Increased observation coverage will improve understanding of the Ocean-Climate System.
Measuring the Ocean: A Plan for Open Source Underwater Robots and Sensors to make Ocean-Science more Accessible	Open-source robots that are simpler in design, cheaper to construct and operate, and carry interoperable and similarly open-source sensors will be more broadly adopted and vastly reduce the barriers to usage and will accelerate the explosion of profiling robots in research, education, and ocean monitoring.
METEOR: A Mobile (Portable) Ocean Robotic Observatory	A paradigm shift in reliable, efficient, near real-time, affordable, integrated data-gathering, assimilating, and ocean modeling to monitor, understand, predict and effectively manage key ocean processes essential to ocean health.

TABLE A.3 Continued

Title	Description
Sustained, Open Access, In-situ, Global Wave Observations for Science and Society	We propose to leverage the NOAA-funded Global Drifter Program (GDP) at Scripps Institution of Oceanography to implement a global array of directional wave spectral drifters (DWSD). Our approach is designed to improve air-sea interaction science and climate assessment. In-situ global wave observations are essential to progress in coupling oceanography, meteorology, and climate science.
Measuring Global Mean Sea Level Changes With Surface Drifting Buoys	We propose to implement a new ocean observing system for monitoring regional and global-mean sea-level rise. This system with consist of a global array of thousands of GPS-driven water-following drifting buoys which will continuously record their geographical positions and the sea levels they ride on.
Super Sites for Advancing Understanding of the Oceanic and Atmospheric Boundary Layers	We propose the concept of "Super Sites" to provide multi-year suites of measurements at specific locations to simultaneously characterize physical and biogeochemical processes within the coupled boundary layers at high spatial and temporal resolution. Measurements will be made from floating platforms, buoys, towers, and autonomous vehicles, utilizing both in situ and remote sensors.
Twilight Zone Observation Network: A Distributed Observation Network for Sustained, Real-time Interrogation of the Ocean's Twilight Zone	The ocean's twilight zone is a vast, globe-spanning region of the ocean. Design and deploy a scalable observation network that, when replicated across the global ocean, will provide data to sustainably manage the oceans twilight zone while conserving the ecosystem services that the zone provides.
Persistent Mobile Ocean Observing: Marine Vehicle Highways	Global infrastructure for Ocean and Planetary Health Monitoring by a fleet of marine vehicles compatible with a standard interface and observe difficult-to-predict transient events that drive ocean processes not accessible to fixed instrumentation. MVH is an opportunity for global partnerships and open to any vehicle developer.
FathomNet: Exploring Our Ocean Using Artificial Intelligence	The adoption of AI in the ocean is limited by the availability of curated data, particularly underwater imagery and video, to train algorithms. FathomNET seeks to automate the processing of underwater imagery and video, to fully explore and discover our ocean.
Battery-free Ocean Internet-of-Thing (IoT)	The Ocean IoT would allow us to monitor aquaculture in realtime, optimize feeding, and other applications. This project seeks to build an underwater IoT using a new backscatter approach, which works better underwater than the traditional approach.
Improved Value of the Observing System through Integrated Satellite and in situ Design	Over the decade this work will demonstrate the value of sustained remotely sensed data to the global ocean observing system and how uncertainties can be overcome through better-informed system design.

continued

TABLE A.3 Continued

Title	Description
A Real-Time Global Rivers Observatory	To understand and document changes in ocean temperatures and mineral sequestering we need to gauge their impact on river basins and the coastal ocean by using the integrative power of fluvial networks to transmit landscape signals. We need to preserve a physical record of changing rivers by building archives of river water and sediment to give future generations a chance to ask novel questions as changes become deeper and more widespread, and to seek answers with new methodologies.
Butterfly: Revealing the Ocean's Impact on Our Weather and Climate	A NASA Earth Venture Mission - EVM3 - Proposed Mission. EVMs are science-driven, competitively selected, low-cost satellite missions.
A Global Network of Surface Platforms for the Observing Air-Sea Interactions Strategy (OASIS)	Air-Sea exchanges of energy, moisture, and gases drive the earth's climate. These interactions influence weather, carbon dioxide distribution, marine life, and human life. OASIS seeks to create a coordinated, multidisciplinary network to observe ocean-atmosphere exchange.
Southern Ocean Storms - Zephyr	We propose a carefully planned addition to the scatterometer constellation to increase the frequency of observations and better constrain high winds. More observations, in conjunction with higher spatial resolution, will reduce the uncertainty in estimates of the global carbon and heat budgets.
UN Ocean Decade Endorsed Actions[a]	
Ocean Biomolecular Observing Network (OBON) (Decade Programme 26)	Ocean life - from viruses to whales - is built from "biomolecules". Biomolecules such as DNA infuse each drop of ocean water, grain of sediment, and breath of ocean air. The Ocean Biomolecular Observing Network (OBON) will develop a global system that will allow science and society to understand ocean life like never before. The programme will transform how we sense, harvest, protect, and manage ocean life, which faces multiple stresses including pollution, habitat loss, and climate change. It will also help communities detect biological hazards like harmful algal blooms and pathogens, and be a key component of next-generation ocean observing systems.

TABLE A.3 Continued

Title	Description
OneArgo: An Integrated Global, Full Depth and Multidisciplinary Ocean Observing Array for Beyond 2020 (Decade Project 114)	OneArgo will transform the revolutionary "core" Argo array (which tracks the upper ocean physical state) to one that has truly global reach, including the polar oceans and marginal seas, extending to the full ocean depth and including ocean biogeochemical measurements. Through Argo's novel data management system, all data will be freely shared in real-time with a high-quality version delivered within 12 months. Implementing OneArgo will greatly increase Argo's already remarkable impact on ocean and climate services, predictions and research, and enable ground-breaking developments in understanding ocean ecosystems, forecasting ocean productivity and constraining the global carbon and energy budgets.
Ocean Decade Research Programme on the Maritime Acoustic Environment (UN-MAE) (Decade Programme 12)	Sound is a persistent yet dynamic component of the maritime environment reflecting both physical and biological properties and phenomenology that define oceanography. Understanding sound in the ocean is critical to support users of, and life within, the ocean. The UN Research Programme on the Maritime Acoustic Environment will establish a comprehensive science-based program aimed at measuring and objectively characterizing underwater acoustic environments - the physical, biological and anthropogenic - at regional to global scales. It will foster new scientific knowledge, technologies, approaches to data collection and dissemination that facilitate the use of sound for analyzing, evaluating and predicting ocean-life systems.
IOGP Environmental Genomics Joint Industry Programme (Decade Contribution 1)	The IOGP Environmental Genomics Joint Industry Programme (eDNA JIP) was launched in June 2019 to coordinate research aimed at exploring the application of eDNA-based analyses in environmental assessments and monitoring of oil and gas offshore and onshore operations. • Oil and gas companies undertake a variety of ecological assessments aimed at characterizing and monitoring the environments in which they operate. • Available conventional approaches for ecological assessment, such as direct sampling and visual/acoustic observation, tend to be more time consuming, expensive, and yield less comprehensive data. • Environmental DNA (eDNA) can be used to detect organisms and estimate biodiversity. This method can help to reduce field time, sampling cost, and be less invasive while often significantly improving the information found by the assessments.

continued

TABLE A.3 Continued

Title	Description
International Ocean Discovery Program (Decade Contribution 140)	The International Ocean Discovery Program (IODP) is an international marine research collaboration that explores Earth's history and dynamics using ocean-going research platforms to recover data recorded in seafloor sediments and rocks and to monitor subseafloor environments. IODP depends on facilities funded by three platform providers with financial contributions from five additional partner agencies. These entities represent twenty-three nations whose scientists are selected to staff IODP expeditions conducted throughout the world's oceans. IODP expeditions are developed from hypothesis-driven science proposals. The program's science plan identifies 14 challenge questions in the four areas of climate change, deep life, planetary dynamics, and geohazards.
IOGP Sound and Marine Life (SML) Joint Industry Programme (JIP) (Decade Contribution 42)	The SML JIP, administered by IOGP, is a partnership of multiple oil and gas companies and the International Association of Geophysical Contractors (IAGC). Established in 2006, the oil and gas exploration and production industry has adopted a pro-active role to improving scientific knowledge and understanding of potential impacts associated with underwater sound, through the establishment of the Sound and Marine Life Joint Industry Programme (SML JIP). Research projects funded through the JIP are categorised into several disciplines: - Sound source characterisation and propagation - Physical effects of sound on the hearing system - Behavioural responses and biological significance - Mitigation and monitoring - Development of research tools.
Deep Ocean Observing Strategy (Decade Programme 129)	DOOS represents an interconnected network of deep-ocean observing, mapping, exploration, and modelling programs working together for the coming decade to 1) characterize the physics, biogeochemistry and biology of the deep ocean in space and time, 2) establish a baseline required to understand changes to its habitats and services, and 3) provide the information needed to have a healthy, predicted, resilient and sustainably-managed (deep) ocean. DOOS will promote the human capital and observing infrastructure needed to address critical scientific and management questions related to the climate, biodiversity and sustainability, while growing a diverse and inclusive next generation of deep-ocean leaders.

TABLE A.3 Continued

Title	Description
The Nippon Foundation-GEBCO Seabed 2030 Project (Decade Programme 107)	Ocean knowledge is critical to understanding our planet yet today we know little about the shape of the ocean floor with 81% yet to be fully mapped. The Nippon Foundation-GEBCO Seabed 2030 Project is a collaboration between The Nippon Foundation and the General Bathymetric Chart of the Oceans (GEBCO) to produce the definitive bathymetric map of the entire ocean by 2030. This is driven by strong motivation to empower the world to make policy decisions, use the ocean sustainably and undertake scientific research informed by detailed understanding of the ocean floor. The map will be freely available for all users.
Promote Seabed 2030 and Ocean Mapping (Decade Contribution 133)	"Promote Seabed 2030 and Ocean Mapping" contributes to the development of a comprehensive digital representation of the ocean. Only 20 percent of the world's ocean is mapped to modern standards, and many parts of the ocean are not surveyed comprehensively with modern multibeam sonar. A full map of the ocean is a crucial starting point for the desired outcomes of the Decade, and failure to produce an adequate map prevents us from globally achieving the "ocean we want."
Global Ocean Biogeochemistry Array (GO-BGC Array) (Decade Contribution 142)	The Global Ocean Biogeochemistry Array (GO-BGC Array) creates a global fleet of robotic floats, transforming how we observe the ocean. The program will release a network of 500 robotic floats into the global ocean to collect chemistry and biology data from the surface to more than 1 mile deep. This program drives a shift in our ability to observe and predict, at the global scale, the effects of climate change on ocean metabolism, carbon uptake, and living marine resource management. Collected data will be freely accessible in near real-time. The program includes an outreach program to diversify the blue workforce.

NOTE: GOOS, Global Ocean Observing System; GPS, global positioning system; IOGP, International Association of Oil & Gas Producers; NASA, National Aeronautics and Space Administration; NOAA, National Oceanic and Atmospheric Administration.

[a] See https://www.oceandecade.org/decade-actions.

TABLE A.4 Connections of the Topical Theme The Restored and Sustainable Ocean to Ocean-Shots and UN Ocean Decade Actions

Title	Description
Ocean-Shots	
The Coral Reef Sentinels: A Mars Shot for Blue Planetary Health	Working alongside NASA AMES and a large international team, this project focuses on more efficient coral reef imaging and analyses. Satellite and drone remote sensing abilities, as well as imaging technology, can be used to get more accurate information about corals, more quickly.
Challenger150: A Global Survey of Deep Sea Ecosystems to Inform Sustainable Management	A blueprint for an inclusive, global, Deep-Sea field program. Emerged from Deep Ocean Stewardship working groups, the project's goal is to coordinate deep ocean research around the globe. With coordination, the project will harness the ability to use ocean knowledge to get us to the Deep Ocean that is wanted.
Twilight Zone Observation Network: A Distributed Observation Network for Sustained, Real-time Interrogation of the Ocean's Twilight Zone	The ocean's twilight zone is a vast, globe-spanning region of the ocean. Design and deploy a scalable observation network that, when replicated across the global ocean, will provide data to sustainably manage the oceans twilight zone while conserving the ecosystem services that the zone provides.
Implementing a Global Deep Ocean Observing Strategy (iDOOS)	The Deep Ocean Observing Strategy (DOOS) is a GOOS project envisioning a globally integrated network of systems that observes the deep ocean (> 200m, with emphasis > 2000m) in support of strong science, policy and planning for sustainable oceans. This ocean shot (iDOOS) will implement an interconnected network of deep-ocean observing, mapping, exploration, and modelling programs working together for the coming decade.
Marine Life 2030: Forecasting Changes to Ocean Biodiversity to Inform Decision-Making - A Critical Role for the Marine Biodiversity Observation Network (MBON)	Marine Life 2030 will establish a globally coordinated system to deliver actionable, transdisciplinary knowledge of ocean life to those who need it, promoting human well-being, sustainable development, and ocean conservation.
Ecological Forecasts for a Rapidly Changing Coastal Ocean	Provide accessible, informative, high-resolution predictions on how changes - from genomes to cells to organisms to ecosystems - may impact people's lives, livelihoods, and property.
Net Ecosystem Improvement: An Evidence-Based Approach	A proposal to continue increasing the size and natural functions of an ecosystem, or natural components of the ecosystem, alongside continuing human development.

TABLE A.4 Continued

Title	Description
Feeding 10 Billion: Contributions from a Marine Circular Bioeconomy	The marine circular bioeconomy will leverage marine aquaculture to sustainably intensify global food production. Nutrient recycling will lead to the co-production of more environmentally favorable, algae-based energy products and materials. Sustainability benefits will include reductions in greenhouse gas emissions, freshwater use, arable land demand, eutrophication, and biodiversity loss.
Future Fisheries in a Changing World	Marine fisheries provide food, income, jobs, and cultural identity for millions of people. Future fisheries face multiple stressors, including climate change. Scientific advances integrating multiple dimensions will be needed to devise effective strategies for climate-resilient fisheries.
Science Enables Abundant Food (SEAFood) with Healthy Oceans	Building SEAFood lifeboat ecosystems that will be able to survive changing ocean temperature and chemistry. Using nutrients from people and agriculture runoff, the effective use of all nutrients could grow 1 billion tons of SEAFood a year.
Transforming Ocean Predictions for Seafood Security and Sustainability (TOPS3)	Understanding interconnected ecosystem observations to forecast seafood security and sustainability.
Meeting Protein & Energy Needs for 10 Billion People While Restoring Oceans	Shellfish and seaweed farming provide resources, opportunities, and solutions to address a wide range of seemingly intractable global problems. Properly placed and managed, aquaculture operations can be restorative to ocean environments, counter climate change, and relieve pressure to farm sensitive terrestrial environments.
The Endless Dive: Marine Species 3D response to climate change in oceans	It is well known throughout different ecological systems that with a warming climate, plant and animal species have moved northward to combat the changing climate. This has also been seen in regards to elevation, from reptilian microhabitats to large-scale mountain ranges. Could marine life follow this pattern? With such a unique, 3D landscape, species can move both deeper and northward in response to climate change. This could lead to impacts on fisheries as well as food chains and communities.
A Call for Health Diagnostics to Preserve Coral Reefs	A set of health diagnostics based on an emerging understanding of the biochemistry of coral reef ecosystems would help mitigate future reef decline and improve remediation efforts of currently compromised reefs by identifying early-stage stress and health decline in corals and other reef organisms.
Plant a Million Corals	The global coral reef crisis requires a scalable solution for communities of varied economies. Plant a Million Corals is an effort to provide affordable access to immediate operations of coral restoration technology.

continued

TABLE A.4 Continued

Title	Description
Reef Solutions: Convergence of Research and Technology to Restore Coral Reefs	A convergence of diverse research expertise and approaches paired with new technological advances are essential to overcoming the challenges associated with rebuilding biodiverse, complex and iconic coral reef ecosystems and ensuring the future of resources utilized by one-eighth of the world's population.
The TeleConnected Reef	Coral reefs are the tropics' most valuable ecosystem. Lack of information about the changing physical and biogeochemical reef environment is a major obstacle to effective efforts. Universal access to real-time and archived model output will enhance predictive capabilities and early warning systems, and transform efforts to manage, conserve, and restore these critical ocean ecosystems in the 21st century.
Seascape Genomics of North Pacific Forage Fishes	We will use a seascape genomics approach to integrate information from multiple disciplines to address large-scale ecosystem and population dynamics as climate conditions continue to change.
Developing Thermally Tolerant Kelp Broodstock to ensure the Global Persistence of Kelp Mariculture in Response to Ocean Change	Transformative work with a holistic approach in which stress memory is used to revitalize wild kelp forests and support the increasing mariculture industry.
PERSEUS (Pelagic Ecosystem Research: Structure, Emergent FUnctions, and Synergies)	A visionary program aims to characterize the composition, connectivity, and complexity of the ocean ecosystem as emergent properties that vary over ecological and evolutionary time skills.
Development of Health Indices for Microbe-Dominated Ocean Systems	Ocean microbes are fundamental to the habitability of Earth. Oceans are changing with climate change, affecting ecosystem processes with unknown consequences. In order to diagnose the health of the oceans, we need to distill oceanographic observations into actionable tools that help us understand and predict the health of the oceans.
Nature-Based Nutrient Reduction for Seagrass Restoration	A three-pillar solution to restore seagrass by addressing natural process bottleneck in shallow waters (<80 meters) and issues with nitrogen levels for seagrass growth. The solutions offer the steps needed to clear a healthy path for new seagrass seedings.

TABLE A.4 Continued

Title	Description
UN Ocean Decade Endorsed Actions[a]	
Deep Ocean Observing Strategy (DOOS) (Decade Programme 129)	DOOS represents an interconnected network of deep-ocean observing, mapping, exploration, and modelling programs working together for the coming decade to 1) characterize the physics, biogeochemistry and biology of the deep ocean in space and time, 2) establish a baseline required to understand changes to its habitats and services, and 3) provide the information needed to have a healthy, predicted, resilient and sustainably-managed (deep) ocean. DOOS will promote the human capital and observing infrastructure needed to address critical scientific and management questions related to the climate, biodiversity and sustainability, while growing a diverse and inclusive next generation of deep-ocean leaders.
Fisheries Strategies for Changing Oceans and Resilient Ecosystems by 2030 (Fish-SCORE 2030) (Decade Programme 63)	Fish-SCORE 2030 will bring together scientists, fishers, resource managers, community practitioners, and policymakers to drive marine fisheries toward climate resilience by 2030. We will develop assessment and modeling frameworks that synthesize complex ecological, social, cultural, economic, and governance dimensions of fishery systems in changing oceans to illuminate specific vulnerabilities and actionable adaptation options. We will delve into evidence and experiences from fishery systems around the globe to find what works in the real world. We will nurture partnerships to apply and improve our frameworks and put to them to work to change the climate outlook of local and regional fisheries.
Global Ecosystem for Ocean Solutions (GEOS) (Decade Programme 172)	GEOS will develop and deploy a series of equitable, durable, and scalable ocean-based solutions for addressing climate change and Ocean Decade's challenges. It will achieve this through three synergistic mechanisms: the GEOS Network made up of researchers, engineers, innovators, investors, decision-makers, and others, which will co-design the GEOS Task Forces for the co-creation of solution-delivering projects, and the GEOS Innovation Engine that will prototype and deploy those solutions. GEOS initial projects focus on ocean-based carbon dioxide removal, providing adaptation tools to coastal communities, and improving ocean-based human health, with further projects to be developed throughout the Ocean Decade.

continued

TABLE A.4 Continued

Title	Description
Ocean Biomolecular Observing Network (OBON) (Decade Programme 26)	Ocean life - from viruses to whales - is built from "biomolecules". Biomolecules such as DNA infuse each drop of ocean water, grain of sediment, and breath of ocean air. The Ocean Biomolecular Observing Network (OBON) will develop a global system that will allow science and society to understand ocean life like never before. The programme will transform how we sense, harvest, protect, and manage ocean life, which faces multiple stresses including pollution, habitat loss, and climate change. It will also help communities detect biological hazards like harmful algal blooms and pathogens, and be a key component of next-generation ocean observing systems.
Sustainability of Marine Ecosystems through global knowledge networks (SMARTNET) (Decade Programme 90)	SMARTNET will establish a global knowledge network (GKN) for ocean science by strengthening and expanding the collaboration of ICES/PICES and partner organizations. It will support and leverage ICES/PICES member countries' activities related to UNDOS, by emphasizing areas of mutual research interest including climate change, fisheries and ecosystem-based management, social, ecological and environmental dynamics of marine systems, coastal communities and human dimensions, and communication and capacity development. It also incorporates strategies to facilitate UNDOS cross-cutting inclusivity themes relating to gender equality, early career engagement, and involvement of indigenous communities and developing nations in the planning and implementation of joint activities.
NOAA Coastal Aquaculture Siting and Sustainability Program (Decade Contribution 51)	The NOAA Coastal Aquaculture Siting and Sustainability (CASS) Program provides high quality science, guidance, and technical support to coastal managers to grow sustainable aquaculture while maintaining and improving ecosystem health. Efforts through the CASS Program are providing needed information to regulatory, industry, and research stakeholders to make sound decisions about permitting, siting, and operating marine fish farms. Continued support of these efforts is guiding monitoring and further research toward ensuring that sustainable practices continue and in minimizing environmental effects. Discharge from marine farms and associated issues of siting such operations are among the most important environmental questions facing this industry.

TABLE A.4 Continued

Title	Description
NSF Coastlines and People (Decade Contribution 135)	The NSF Coastlines and People (CoPe) program supports diverse, innovative, multi-institution Coastal Research Hubs that are focused on critically important coastlines and people research that is integrated with broadening participation goals. The hubs are structured using a convergent science approach, at the nexus between coastal sustainability, human dimensions, and coastal processes to transform understanding of interactions among natural, human-built, and social systems in coastal, populated environments. CoPe supports Focused Hubs, projects $1 million or less per year for 3 to 5 years, as well as Large-scale Hubs, projects of $2-4 million per year, for up to 5 years.
SUstainability, Predictability and REsilience of Marine Ecosystems (SUPREME) (Decade Programme 118)	Changing oceans are significantly impacting valuable marine species and the many people, communities, and economies that depend upon them. Warming oceans, rising seas, decreasing ocean ice, increasing ocean acidification, and extreme events (e.g., marine heat waves) are affecting the distribution and abundance of marine species in many regions. These changes are expected to increase with continued climate change and there is much at risk. The SUPREME programme seeks to globally implement an infrastructure to support robust climate- and ocean-related forecasts, predictions, and projections to guide marine ecosystem management and adaptation strategies that reduce risks and increase resilience of marine/coastal resources and the people who depend on them.
Coral Reef Restoration Engaging Local Stakeholders Using Novel Biomimicking IntelliReefs (Decade Project 112)	In November 2018, IntelliReefs deployed three nanotechnology artificial reefs underwater near Philipsburg, Sint Maarten to restore coral reefs after Hurricane Irma (2017). IntelliReefs mimic natural, healthy reefs and provide bioengineered habitat for faster growing and more resilient reefs. IntelliReefs are designed down to the nanoscale for site, species and function, allowing for immediate marine integration. IntelliReefs provide food and shelter for fish populations, attract wild corals, rapidly grow healthy coral reef communities, and enhance local biodiversity. Over the next decade, we will deploy additional IntelliReefs and further analyze the benefits and applications of IntelliReefs for fish, corals, and local economies.

NOTE: GOOS, Global Ocean Observing System; ICES, International Council for the Exploration of the Sea; NASA, National Aeronautics and Space Administration; NOAA, National Oceanic and Atmospheric Administration; NSF, National Science Foundation; PICES, The North Pacific Marine Science Organization; UNDOS, United Nations Decade on Ocean Science and Sustainable Development.

[a] See https://www.oceandecade.org/decade-actions.

TABLE A.5 Connections of the Topical Theme Ocean Solutions for Climate Resilience to Ocean-Shots and UN Ocean Decade Actions

Title	Description
Ocean-Shots	
Butterfly: Revealing the Ocean's Impact on Our Weather and Climate	A NASA Earth Venture Mission - EVM3 - Proposed Mission. EVMs are science-driven, competitively selected, low-cost satellite missions.
Southern Ocean Storms - Zephyr	We propose a carefully planned addition to the scatterometer constellation to increase the frequency of observations and better constrain high winds. More observations, in conjunction with higher spatial resolution, will reduce the uncertainty in estimates of the global carbon and heat budgets.
Observing the Oceans Acoustically	Our vision is a global-spanning multi-purpose ocean acoustic network in direct analogy with GPS required to transform use and observation of the world below the ocean surface. A small number of judiciously-placed low-frequency acoustic sources transmitting to globally distributed receivers enable monitoring of acoustic geo-positioning, ocean measurements, and ocean sound time series.
SMART Subsea Cables for Observing the Ocean and Earth	SMART Cables' vision is to observe the oceans and Earth with a planetary-scale network of sensor-enabled submarine telecommunications cables to support climate, ocean circulation, sea-level monitoring, and tsunami and earthquake early warning and disaster risk reduction.
Carbon Sequestration via Drilling-Promoted Seawater-Rock Interactions	Finding effective methods to limit the accumulation of atmospheric CO_2 through sequestration encouraged by seawater-rock interactions.
Caribbean Observatories (CARIBO): Ocean Storminess at the Western Boundary and Its Impacts on Shelf/Slope Environment and Ecosystems	Innovative, multi-disciplinary, multi-scale, observations at the inflow and outflow of the Caribbean Seas, one of the ocean's most biologically diverse ecosystems serving 38 countries/dependencies with large inequalities in governance and wealth.
Measuring Global Mean Sea Level Changes With Surface Drifting Buoys	We propose to implement a new ocean observing system for monitoring regional and global-mean sea-level rise. This system with consist of a global array of thousands of GPS-driven water-following drifting buoys which will continuously record their geographical positions and the sea levels they ride on.
A Global Network of Surface Platforms for the Observing Air-Sea Interactions Strategy (OASIS)	Air-Sea exchanges of energy, moisture, and gases drive the earth's climate. These interactions influence weather, carbon dioxide distribution, marine life, and human life. OASIS seeks to create a coordinated, multidisciplinary network to observe ocean-atmosphere exchange.

TABLE A.5 Continued

Title	Description
Navigating the Ocean's Role in Carbon Dioxide Removal	We propose an Ocean Shot to develop the science we need to assess the ocean's ability to sequester atmospheric CO_2 and understand how the ocean can intentionally and responsibly be modified to increase uptake of atmospheric CO_2.
A Real-Time Global Rivers Observatory	To understand and document changes in ocean temperatures and mineral sequestering we need to gauge their impact on river basins and the coastal ocean by using the integrative power of fluvial networks to transmit landscape signals. We need to preserve a physical record of changing rivers by building archives of river water and sediment to give future generations a chance to ask novel questions as changes become deeper and more widespread, and to seek answers with new methodologies.
Super Sites for Advancing Understanding of the Oceanic and Atmospheric Boundary Layers	We propose the concept of "Super Sites" to provide multi-year suites of measurements at specific locations to simultaneously characterize physical and biogeochemical processes within the coupled boundary layers at high spatial and temporal resolution. Measurements will be made from floating platforms, buoys, towers, and autonomous vehicles, utilizing both in situ and remote sensors.
Mining Five Centuries of Climate and Maritime Weather Data from Historic Records	Building on established protocols from prior data rescue efforts to transfer qualitative information to quantitative weather records, digitizing and extracting historic observations greatly expands current records of maritime weather around the world back to ~1500 CE, feeds into community reanalysis modeling efforts, and offers the opportunity to engage the public through citizen science.
Why Paleoceanographic Observations are Needed to Improve Future Climate Projections	Ocean heat uptake is an important determinant of the earth's climate sensitivity to atmospheric carbon dioxide. A proposal about breaking the impasse in improving climate sensitivity estimates by considering instrumental and proxy evidence simultaneously.
OceanPredict.US	Leveraging national and international collaborations and partnerships, OceanPredict.US aims to coalesce national global and basin-scale ocean assessment and prediction capabilities into operational implementations supporting nested regional and coastal activities. OceanPredict.US targets the full operational oceanography value chain, spanning: needs, observations, data management, analysis, ocean prediction, dissemination via a digital ocean, and information service delivery, culminating in end use while enabling innovation and value-added development.

continued

TABLE A.5 Continued

Title	Description
A Sensor Network for Mixing at the Ocean's Bottom Boundary	A community-level project gathering observationalists, turbulence modelers, and general circulation model users and builders. The project core will be the development of a sensor network to estimate momentum and buoyancy fluxes within the planetary boundary layer, stress-driven drag at the bottom boundary, and energy conversion rates associated with the flow over topography.
UN Ocean Decade Endorsed Actions[a]	
Global Ecosystem for Ocean Solutions (GEOS) (Decade Programme 172)	GEOS will develop and deploy a series of equitable, durable, and scalable ocean-based solutions for addressing climate change and Ocean Decade's challenges. It will achieve this through three synergistic mechanisms: the GEOS Network made up of researchers, engineers, innovators, investors, decision-makers, and others, which will co-design the GEOS Task Forces for the co-creation of solution-delivering projects, and the GEOS Innovation Engine that will prototype and deploy those solutions. GEOS initial projects focus on ocean-based carbon dioxide removal, providing adaptation tools to coastal communities, and improving ocean-based human health, with further projects to be developed throughout the Ocean Decade.
Observing Air-Sea Interactions Strategy (OASIS) (Decade Programme 97)	Air-sea exchanges of energy, moisture, and gases drive and modulate the Earth's weather and climate, influencing life, including our own. These air-sea interactions fuel the hydrological cycle and affect precipitation across the globe. Air-sea interactions affect the distribution of carbon dioxide between the atmosphere and ocean, how seawater flows and winds blow, and how pollutants floating on the ocean surface move - information critical to policymakers, industry, and civil society. The Observing Air-Sea Interactions Strategy (OASIS) PROGRAMME will provide observational-based knowledge to fundamentally improve weather, climate and ocean prediction, promote healthy oceans, the blue economy, and sustainable food and energy.

TABLE A.5 Continued

Title	Description
Blue Climate Initiative - Solutions for People, Ocean, Planet (Decade Programme 138)	Human health and well-being depend upon a healthy ocean for needs as diverse as food, oxygen, a stable climate, moderate weather and livelihoods - and a well-managed and thriving ocean can greatly contribute to improved human health and well-being. The Blue Climate Initiative brings together scientists, community groups, engineers, entrepreneurs, investors, government leaders and global influencers to collaboratively identify, develop and implement science-based programs to protect the ocean and use the ocean's remarkable power and potential to tackle climate change and other urgent issues of our time, from renewable energy and sustainable food supplies to human health and resilient ocean economies.
A Transformative Decade for the Global Ocean Acidification Observing System (Decade Contribution 116)	Ocean acidification (OA) is the ongoing observed increase in seawater acidity (pH) primarily due to the ocean's uptake of anthropogenic atmospheric carbon dioxide (CO_2). The rate of the ocean's changing chemistry is measured by a suite of stations worldwide, and conditions expected by 2100 will have several negative effects on marine life. Many challenges to understanding OA and its impacts remain. A robust understanding of OA and its impacts requires interdisciplinary monitoring and research efforts, including carbonate chemistry, physical oceanography, biogeochemistry, ecology, biology, natural resource economics, and other social and hard sciences. It also requires a global workforce that can analyze, assessing, and applying this data. This Decade Programme expands CO_2 observing systems by developing the next generation of sensors, training new experts, ensuring materials are available for accurate measurements, and filling in under-observed regions. It also builds capacity for publicly-available data that is fed into products useful for stakeholders.
NASA Sea Level Change Science Team (Decade Contribution 33)	Since 2014, the NASA Sea Level Change Science Team (N-SLCT) has been conducting interdisciplinary sea level science by collecting and analyzing observational evidence of sea level change, quantifying underlying causes and driving mechanisms, producing projections of future changes in sea level, and communicating NASA's latest discoveries to the public through the Sea Level Portal at https://sealevel.nasa.gov. N-SLCT has made progress many important problems in sea level science, resulting in a better understanding of ice sheet dynamics, ocean processes, the development of tools and assessments of mass loss impacts from ice sheets and glaciers on coastal cities, and improved representation of vertical land motion related to coastal subsidence, tectonics, and Earth's post-glacial rebound.

continued

TABLE A.5 Continued

Title	Description
ForeSea - The Ocean Prediction Capacity of the Future (Decade Programme 28)	ForeSea's vision is for strong international coordination and community building of an ocean prediction capacity for the future. The overarching goal are to (1) improve the science, capacity, efficacy, use, and impact of ocean prediction systems and (2) build a seamless ocean information value chain, from observations to end users, for economic and societal benefit. These transformative goals aim to make ocean prediction science more impactful and relevant.
CoastPredict - Observing and Predicting the Global Coastal Ocean (Decade Programme 144)	CoastPredict will transform the science of observing and predicting the Global Coastal Ocean, from river catchments, including urban scales, to the oceanic slope waters. It will integrate observations with numerical models to produce predictions with uncertainties from extreme events to climate, for the coastal marine ecosystems (their services), biodiversity, co-designing transformative response to science and societal needs. CoastPredict will re-define the concept of the Global Coastal Ocean, focusing on the many common worldwide features, to produce observations and predictions of natural variability and human-induced changes in the coastal areas and upgrade the infrastructure for exchange of data with standard protocols.
Science Monitoring And Reliable Telecommunications (SMART) Subsea Cables: Observing the Global Ocean for Climate Monitoring and Disaster Risk Reduction (Decade Project 94)	The Joint Task Force for Science Monitoring And Reliable Telecommunications (JTF SMART) subsea cables is facilitating the integration of environmental sensors into trans-ocean commercial submarine telecommunications cables toward a planetary scale array that monitors ocean climate and sea level rise. The network will revolutionise real-time warning systems for earthquake and tsunami disaster mitigation. The first major SMART project is underway in Portugal, with others in various stages of planning and funding - "Big Tech Blue Economy. The JTF will provide coordination between all stakeholders while catalyzing education, training and outreach programs to build capacity and societal benefit."

NOTE: GPS, global positioning system; NASA, National Aeronautics and Space Administration.
[a] See https://www.oceandecade.org/decade-actions.

TABLE A.6 Connections of the Topical Theme Healthy Urban Seas to Ocean-Shots and UN Ocean Decade Actions

Title	Description
Ocean-Shots	
EquiSea: The Ocean Science Fund for All	EquiSea aims to improve equity in ocean science by establishing a philanthropic fund to provide direct financial support to projects, coordinating capacity development activities, fostering collaboration and co-financing of ocean science between academia, government, NGOs, and private sector actors, and supporting the development of low-cost and easy-to-maintain ocean science technologies.
Feeding 10 Billion: Contributions from a Marine Circular Bioeconomy	The marine circular bioeconomy will leverage marine aquaculture to sustainably intensify global food production. Nutrient recycling will lead to the co-production of more environmentally favorable, algae-based energy products and materials. Sustainability benefits will include reductions in greenhouse gas emissions, freshwater use, arable land demand, eutrophication, and biodiversity loss.
Global Ocean and Human Health Program	This Ocean Shot addresses the need for an international program to meet the challenges at the intersection of ocean health and human health. A Global Ocean and Human Health (GOHH) Program would build a transformative network encompassing essential research and engineering, policy, and economic concerns relevant to ocean and human health in the U.S. and globally.
TRITON: A Social Media Network for the Ocean	A different, unified approach to outreach and communication: TRITON, a social media platform. The platform will serve as a one-stop-shop for information about the ocean and society, a place where scientists and organizations can share and amplify science-based information and for dissemination of original content that connects people to the ocean through the lens of their own community experiences.
OceanCloud: Transforming Oceanography with a New Approach to Data and Computing	OceanCloud is a new vision for data infrastructure with three pillars: 1. Analysis-ready data shared openly in the cloud 2. On-demand, scalable distributed processing 3. Interactive, data-proximate computing, available to anyone
Marine Health Hubs: Building Interdisciplinary Regional Hubs of Excellence to Research and Address the Societal Impacts of Marine Debris Across the Globe	Marine Health Hubs (MHH) program will build capacity for and mobilize interdisciplinary teams worldwide to address marine debris across the continuum of research to application. We will establish self-sustaining regional hubs of excellence to promote interdisciplinary collaboration to tackle plastic marine waste from an environmental, economic, and societal lens.

continued

TABLE A.6 Continued

Title	Description
An Ocean Science Education Network for the Decade	A coordinated ocean science education network will leverage resources across sectors and nations to reach global citizenry. A flow of information among scientists, education professionals, indigenous leaders, policymakers, business leaders, and the public will help guide research priorities and enhance global ocean literacy.
Novel Coastal Ecosystems: Engineered Solutions to Accelerate Water Quality Restoration using Engineered Aeration	Low-oxygen conditions occurring in coastal waters are increasingly driven by nutrient release from human activities impairing ecosystems. Climate changes and ecosystem regime shifts have altered the baselines used to generate restoration targets, so remediation trajectories are now uncertain and management efforts may fall short. A solution involves augmenting nutrient reduction efforts with engineered aeration to make up the difference. Aeration of the water column could not only reverse hypoxia but also accelerates processes that naturally remove nutrients. Thoughtful studies of ecosystem function under engineered conditions are needed to determine feasibility of these solutions and document potential unintended consequences.
OceanPredict.US	Leveraging national and international collaborations and partnerships, OceanPredict.US aims to coalesce national global and basin-scale ocean assessment and prediction capabilities into operational implementations supporting nested regional and coastal activities. OceanPredict.US targets the full operational oceanography value chain, spanning: needs, observations, data management, analysis, ocean prediction, dissemination via a digital ocean, and information service delivery, culminating in end use while enabling innovation and value-added development.
An INFOstructure solution to the socio-ecological hazards of coastal flood control infrastructure	Our Ocean Shot concept is towards creating an information structure (Info-structure) to support decision making surrounding the use of engineering infrastructure for flood hazards in coastal areas. Often, use of such infrastructure occurs in the absence of socio-ecological data about coastal ecosystems and communities and result in unforeseen and multifaceted coastal disasters. The concept will be framed around case studies in the U.S. Gulf South and The Netherlands that highlight the increasing and changing use of flood control structures in response to climate impacts. The concept is scalable other ecosystems and geographies facing climate change related flooding.

TABLE A.6 Continued

Title	Description
Revolutionizing Coastal Ocean Research through a Novel Share Model for the Long-term Sustainability of Humanity	We propose a bold vision for conducting ocean science. While the coastal ocean is essential to the well-being and long-term sustainability of humanity, our scientific knowledge on coastal ecosystems falls woefully short. We are calling for a new model that addresses social, economic, political, and logistical factors that impede participation in coastal ocean science through instruments and community relationships. Our new model will build on familiar tenets of the share economy developed to respond to the needs of coastal ocean research and science.
UN Ocean Decade Endorsed Actions[a]	
Fisheries Strategies for Changing Oceans and Resilient Ecosystems by 2030 (Decade Programme 63)	Fish-SCORE 2030 will bring together scientists, fishers, resource managers, community practitioners, and policymakers to drive marine fisheries toward climate resilience by 2030. We will develop assessment and modeling frameworks that synthesize complex ecological, social, cultural, economic, and governance dimensions of fishery systems in changing oceans to illuminate specific vulnerabilities and actionable adaptation options. We will delve into evidence and experiences from fishery systems around the globe to find what works in the real world. We will nurture partnerships to apply and improve our frameworks and put to them to work to change the climate outlook of local and regional fisheries.
NSF Coastlines and People (Decade Contribution 135)	The NSF Coastlines and People (CoPe) program supports diverse, innovative, multi-institution Coastal Research Hubs that are focused on critically important coastlines and people research that is integrated with broadening participation goals. The hubs are structured using a convergent science approach, at the nexus between coastal sustainability, human dimensions, and coastal processes to transform understanding of interactions among natural, human-built, and social systems in coastal, populated environments. CoPe supports Focused Hubs, projects $1 million or less per year for 3 to 5 years, as well as Large-scale Hubs, projects of $2-4 million per year, for up to 5 years.

continued

TABLE A.6 Continued

Title	Description
Estuarine Ecological Knowledge Network (EEKN) (Decade Project 43)	The Estuarine Ecological Knowledge Network (EEKN) is designed to utilize the everyday experiences and traditional ecological knowledge of coastal community members in informing scientific research and policy decisions regarding estuarine environments which are crucial in maintaining the health and productivity of Earth's oceans. The EEKN will create a dispersed network of community observers and "citizen scientists," who will document the ecological conditions that they encounter, providing data for ocean scientists and feedback to policy makers. The EEKN will also document and preserve the knowledge of indigenous peoples and encourage the participation of under-represented groups in marine science and management.
An Ocean Corps for Ocean Science (Decade Programme 9)	Motivated by the example of the US Peace Corps, we propose "An Ocean Corps for Ocean Science" as a unifying concept for sustaining long-term education and research collaborations between scientists from under-resourced nations and higher-resourced nations. Based upon our experience in Peace Corps and with the Coastal Ocean Environment Summer School in Ghana (https://coessing.org), we are confident that an Ocean Corps would inspire large numbers of scientists, especially early-career scientists, into its ranks, thus "internationalizing" their outlook, molding many of them into champions for international capacity development for the remainder of their careers, and fostering true ocean science collaborations worldwide.
A multi-dimensional and inclusive approach for transformative capacity development (CAP-DEV 4 the Ocean) (Decade Project 39)	This initiative brings together a diverse multi-national partnership to address the critical need for capacity development in ocean science, literacy and conservation in low- and middle-income countries. We seek to strengthen systems (funding, mentorship, recognition, learning, communication and evaluation) that address the challenges of the Ocean Decade and support the capacity to deliver ocean science and leadership, which is central to healthy oceans and sustainable development. This will proactively advance the integration of capacity development in the design and implementation of priorities for the Ocean Decade.

NOTE: NGO, nongovernmental organization; NSF, National Science Foundation.
[a] See https://www.oceandecade.org/decade-actions.

Appendix B

Committee Biographies

Larry A. Mayer (NAE) (*Chair*) is the director of the School of Marine Science and Ocean Engineering and the Center for Coastal and Ocean Mapping, the co-director of the Joint Hydrographic Center, and a professor of Earth science and ocean engineering at the University of New Hampshire. His research interests include sonar imaging, remote characterization of the seafloor, and advanced applications of three-dimensional visualization to ocean mapping challenges. Dr. Mayer received his Ph.D. from the Scripps Institution of Oceanography in marine geophysics in 1979, and graduated magna cum laude with an Honors degree in geology from the University of Rhode Island in 1973. At Scripps his future path was determined when he worked with the Marine Physical Laboratory's Deep-Tow Geophysical package, but he applied this sophisticated acoustic sensor to study the history of climate. Dr. Mayer has participated in more than 90 cruises and has been the chief or the co-chief scientist of numerous expeditions, including two legs of the Ocean Drilling Program. He has served on and chaired many international panels and committees and has the requisite large number of publications on a variety of topics in marine geology and geophysics. He is the recipient of the Keen Medal for Marine Geology, an Honorary Doctorate from the University of Stockholm, and the University of New Hampshire's and the University of Rhode Island's Graduate School of Oceanography's Distinguished Alumni Award. Dr. Mayer served on the President's Panel for Ocean Exploration and chaired the 2004 National Research Council's Committee on National Needs for Coastal Mapping and Charting. In 2013, he chaired the National Academies of Sciences, Engineering, and Medicine's Committee on the Impacts of *Deepwater Horizon* on the Ecosystem Services of the Gulf of Mexico. He became a member of the National Academy of Engineering in 2018. Dr. Mayer is the chair of the Ocean Studies Board at the National Academies.

Mark R. Abbott (*Vice Chair*) was the 10th director and president of Woods Hole Oceanographic Institution. Dr. Abbott was the dean of the College of Earth, Ocean, and Atmospheric Sciences at Oregon State University. He was an investigator in the Office of Naval Research's (ONR's) Coastal Transition Zone program and the Eastern Boundary Current program. He is presently funded by ONR to explore advanced computer architectures for use in undersea platforms. Dr. Abbott has also advised ONR and the National Science Foundation (NSF) on issues regarding advanced computer technology and oceanography. He was also a member of MEDEA, which advised the federal government on issues of national security and climate change. In 2006, Dr. Abbott was appointed by the president to a 6-year term on the National Science Board, which oversees NSF and provides scientific advice to the White House and to Congress. Dr. Abbott was appointed in 2008 by Oregon Governor Kulongoski as the vice chair of the Oregon Global Warming Commission, which is leading the state's efforts in mitigation and adaptation strategies in response to climate change. In 2011, Dr. Abbott was the recipient of the Jim Gray eScience Award, presented by Microsoft Research to a nationally recognized researcher who has made outstanding contributions to data-intensive computing. He served as the president of The Oceanography Society from 2013 to 2014. He also served on the Board of Trustees for the University Corporation for Atmospheric Research and the Consortium for Ocean Leadership. Dr. Abbott holds a B.S. in conservation of natural resources from the University of California, Berkeley, and a Ph.D. in ecology from the University of California, Davis. Dr. Abbott is a member of the Ocean Studies Board at the National Academies.

Carol Arnosti is a professor in the Department of Marine Sciences at the University of North Carolina at Chapel Hill. Dr. Arnosti is a marine chemist who works at the intersection of chemistry, microbiology, and molecular ecology to investigate the relationships between the structure of organic matter and the rates and pathways by which it is degraded by microbial communities. She has pioneered new methods to measure the activities of the extracellular enzymes that initiate microbially driven carbon cycling; she has used these methods to develop a mechanistic understanding of the variable reactivity of marine organic matter and to reveal structure-function relationships within organic-matter processing microbial communities. Fieldwork for these studies has taken place in the Atlantic, Pacific, and Arctic Oceans, and has resulted in large-scale oceanic transects revealing changing rates and substrate patterns in microbial organic matter utilization. Her extensive research collaborations in Germany have been facilitated by Fulbright and Hanse Fellowships. Dr. Arnosti received a B.A. in chemistry from Lawrence University and a Ph.D. in chemical oceanography from the Massachusetts Institute of Technology/Woods Hole Oceanographic Institution Joint Program in Oceanography. Dr. Arnosti is a member of the Ocean Studies Board at the National Academies.

Claudia Benitez-Nelson is the associate dean for instruction, community engagement, & research and the Carolina Distinguished Professor and Endowed Chair in Marine Studies in the College of Arts & Sciences at the University of South Carolina. As an associate dean, Dr. Benitez-Nelson has direct oversight of five departments (Biological Sciences; Mathematics; School of the Earth, Ocean & Environment; Psychology; and Statistics) that encompass more than 300 faculty and staff, 300 graduate students, and more than 2,500 undergraduate majors. Dr. Benitez-Nelson's research focuses on the biogeochemical cycling of phosphorus and carbon and how these elements are influenced by both natural and anthropogenic processes. She is a diverse scientist, with expertise ranging from radiochemistry to harmful algal bloom toxins and is highly regarded for her cross-disciplinary research. Over the past two decades, Dr. Benitez-Nelson has authored or co-authored more than 100 articles, including lead author publications in the journals *Science* and *Nature*. She has been continuously supported by substantial, multi-year research and education grants from the National Science Foundation (NSF) and the National Aeronautics and Space Administration, among others. Her many research honors include the Early Career Award in Oceanography from the American Geophysical Union (AGU) and Fulbright and Marie Curie Fellowships. In 2015 she was named an American Association for the Advancement of Science Fellow, and in 2017, was named an Association for the Sciences of Limnology and Oceanography Sustaining Fellow. Dr. Benitez-Nelson is also highly regarded as a teacher and mentor, having received the National Faculty of the Year Award from the National Society of Collegiate Scholars and the University of South Carolina's Distinguished Professor of the Year Award, SEC Faculty Achievement Award, and Outstanding Faculty Advisor of the Year. In 2014, she received the Sulzman Award for Excellence in Education and Mentoring from the Biogeosciences Section of AGU. Dr. Benitez-Nelson is regularly called on by national and international scientific and policy agencies for her expertise and currently serves or has served as a member of the Advisory Committee to the Geoscience Directorate of NSF, the U.S. Environmental Protection Agency's Science Advisory Board, and the National Academies' Ocean Studies Board. Dr. Benitez-Nelson earned a B.S. in chemistry and oceanography from the University of Washington and a Ph.D. in oceanography from the Woods Hole Oceanographic Institution/Massachusetts Institute of Technology Joint Program in 1999.

Anjali Boyd is a Duke University Ph.D. student and a Dean's Graduate Fellow in the Nicholas School of the Environment. She received her B.S. in marine science from Eckerd College as a National Oceanic and Atmospheric Administration Hollings Scholar. Her research examines how species interactions (both intra and inter) and physical forces interact to regulate the recovery of foundation species to environmental stressors. Boyd aspires to develop novel ecosystem-based restoration and management practices to restore foundation species worldwide. She is fiercely committed to increasing representation of women and ethnic minorities in ocean sciences and elevating the voices and contributions of student and early-career scientists. Thus far, she has served as the chair and the vice chair of the Ecological Society of America's Student Section; an appointed member of the Ecological Society of America's Diversity, Equity, Inclusion, and Justice taskforce; the treasurer/secretary of the Society of Wetland Scientists' Student Section; and a member of the Coastal and Estuarine Research Federation Broadening Participation Committee. Additionally, as the director of iNviTechnology, she works to combat the underrepresentation of women and ethnic minorities in science, technology, engineering, and mathematics fields through educational entrepreneurial programs to engage young children ages 0–5 and K–12 students. Boyd also serves as an elected official in her hometown of Durham, North Carolina, as the Durham County Soil and Water Conservation District Supervisor.

Annie Brett is an assistant professor at the University of Florida Levin College of Law, where she teaches and writes on ocean and coastal law and the intersection of law and science. Her scholarship focuses on how scientific data are used in environmental decision-making, including data collected using emerging methods and technologies. In addition to legal venues, Dr. Brett has published in leading scientific outlets, including *Nature*, and presented in national and international policy forums. Prior to joining the University of Florida Levin College of Law, Dr. Brett worked on international ocean policy for the Stanford University Center for Ocean Solutions and the World Economic Forum. She is an accomplished mariner, recognized as the youngest female vessel captain to operate in the Pacific. Dr. Brett received her A.B. from Harvard University and J.D. and Ph.D. from the University of Miami.

Thomas S. Chance served as the chief executive officer (CEO) and board chairman of ASV Global, an international leader in autonomous surface vehicle technology (ASV) for 9 years. Since 2010, when Mr. Chance founded the company, ASV Global has built more than 90 unmanned vessels and integrated 30 different ocean science payload types for defense, commercial, and academic applications. Mr. Chance's role was product strategy and development. He also founded C&C Technologies in 1992, a global leader in autonomous underwater vehicle (AUV) technology where he was the CEO and the board chairman. C&C was the first company in the world to offer AUV survey services to the offshore oil industry and has since amassed almost a half million line kilometers of actively propelled AUV data collection. His other work includes maritime technology development for the U.S. Naval Research Laboratory, the Office of Naval Research, and hydrographic survey operations in shallow and deep waters across the globe.

With 550 employees and 10 offices worldwide, he sold C&C Technologies to Oceaneering International in early 2015. Earlier, he was the vice president of business and systems development at Fugro Chance. He was a member of the National Research Council's Committee on National Needs for Coastal Mapping and Charting. He has given dozens of conference presentations over the past 30 years. He was the chair of the National Ocean Industries Association Technology Policy Committee. He has received numerous awards including the BP Upstream Innovation Award, the Marine Technology Corporate Excellence Award, Innov8 Acadiana Award, 1996 Entrepreneur of the Year for the Gulf Coast Region, 2006 LCG International Achievement Award, 2008 Marine Technology Reporter Seamaster of the Year Award, and 2015 Junior Achievement Business Person of the Year Award. He earned an M.S. in engineering and industrial management at Purdue University. He retired as the CEO of ASV Global in May 2019 and is now working as an independent consultant. Mr. Chance is a member of the Ocean Studies Board at the National Academies.

Daniel Costa is the director of the Institute of Marine Sciences and the Distinguished Professor of Ecology and Evolutionary Biology at the University of California, Santa Cruz. Dr. Costa completed a B.A. at the University of California, Los Angeles, and a Ph.D. at the University of California, Santa Cruz, followed by postdoctoral research at the Scripps Institution of Oceanography. His research focuses on the ecology and physiology of marine mammals and seabirds, taking him to every continent and almost every habitat from the Galapagos to Antarctica. He has worked with a broad range of animals including turtles, penguins, albatross, seals, sea lions, sirenians, whales, and dolphins and has published more than 400 scientific papers. His current work is aimed at recording the movement and distribution patterns of marine mammals and seabirds in an effort to understand their habitat needs. This work is helping to identify biodiversity hot spots and the factors that create them. He has been developing tools to identify and create viable Marine Protected Areas for the conservation of highly migratory species. In addition, his research is studying the response of marine mammals to underwater sounds and developing ways to assess whether the potential disturbance may result in a population consequence. He has been active in graduate education having supervised 22 master's students, 30 doctoral students, and 15 postdoctoral scholars. With Barbara Block he co-founded the Tagging of Pacific Predators program, a multidisciplinary effort to study the movement patterns of 23 species of marine vertebrate predators in the North Pacific Ocean. He is an internationally recognized authority on tracking of marine mammals and birds. He has served as a member of a number of international science steering committees including the Integrated Climate and Ecosystem Dynamics program, The Census of Marine Life, Southern Ocean Global Ocean Ecosystems Dynamics, Climate Impacts on Top Predators, the Southern Ocean Observing System, and the Integrated Marine Biogeochemistry and Ecosystem Research. Dr. Costa is a member of the Ocean Studies Board at the National Academies.

John R. Delaney is a professor emeritus in oceanography at the University of Washington. Early career experiences involved working as an economic geologist across the western United States searching for base and precious metal deposits, and living for 6 months on and within multiple Galapagos volcanoes in the early 1970s. With the blessings of the National Academies of Sciences, Engineering, and Medicine's Ocean Studies Board, in 1987 Dr. Delaney initiated the National Science Foundation (NSF)-sponsored RIDGE Program by working with many colleagues to focus on Mid-Ocean Ridge (MOR) systems, from the perspective of the feedback loops involving physical, chemical, and biological processes exposed within highly dynamic volcano-hydrothermal systems on the seafloor. RIDGE quickly expanded into "Inter-RIDGE" (international), which continues today and is being managed by the French at this time. Having made many tens of dives with the deep submersible ALVIN, and with nearly 50 sea-going expeditions, Dr. Delaney, again working with colleagues, helped launch an effort focused on gaining long-term, sustained access to myriad processes operating on, below, and above the seafloor off the coasts of Washington and Oregon. This effort involved implementing a tectonic plate–scale installation of a submarine, electro-optically cabled network of sensors across the Juan de Fuca (JdF) Plate as part of the NSF-Ocean Observatory Initiative. As of 2020, that cabled array has been operating for half a decade and provides continuous real-time data and information to researchers and educators alike about many processes operating both along the JdF Ridge and within the Cascadia Subduction Zone. It is also documenting a host of water-column processes off shore. An important future use of this system will be to capture, for the first time, in real-time, the entire process of a MOR eruption using the cable-power to operate and interact with remote "Resident" Autonomous Undersea Vehicles at the seafloor and in the overlying ocean. In addition to studying the linkages among volcanoes and life here on Earth, Dr. Delaney has also been interested in the search for life elsewhere in the solar system. He worked with National Aeronautics and Space Administration-Jet Propulsion Laboratory to help plan space missions to the moons of Jupiter. He is a fellow of the American Geophysical Union (AGU) and the recipient of AGU's Athelstan Spilhaus Award for fostering public engagement by conveying to the general public the excitement, significance, and beauty of Earth and Space science. A late-stage focus of Dr. Delaney's career has become the "Global Ocean-Human Culture" array of themes involving public elaboration of the broad, interlinked spectrum of ways that human societies have used or been impacted by oceans for tens of millennia, and how we may be impacted for centuries to come. The goal is to fundamentally expand public awareness about how crucial the ocean-systems are to the long-term well-being of all humanity from environmental, economic, security, and discovery perspectives. Dr. Delaney is a member of the Ocean Studies Board at the National Academies.

Angee Doerr is an assistant professor of practice and fisheries extension specialist with Oregon Sea Grant and Oregon State University (OSU). Prior to this, she spent several years as a research scientist with Stanford's Center for Ocean Solutions. At OSU, Dr. Doerr focuses on fisheries and other marine coastal resources, providing community outreach and education on a range of subjects, to include marine resource management, aquaculture, climate impacts, nearshore energy, and sustainable economic growth for coastal industries. Dr. Doerr works closely with community partners, including commercial fishers, managers, and researchers, to advance our understanding of fisheries and other marine resources in Oregon and along the U.S. West Coast. She has a B.Sci. and an M.B.A., as well as a Ph.D. focused on socially and ecologically sustainable fisheries. Dr. Doerr is currently a Commander in the U.S. Navy Reserves, having spent 8 years on active duty as a Naval Flight Officer and 10 years in the Reserves as the Officer-in-Charge of a variety of units.

Scott Glenn is a distinguished professor in the Department of Marine and Coastal Sciences at Rutgers University and the co-director of the Center for Ocean Observing Leadership. Dr. Glenn graduated from the Massachusetts Institute of Technology and Woods Hole Oceanographic Institution Joint Program in 1983 with an Sc.D. in ocean engineering. His more than 35-year research career of developing and implementing sustained real-time ocean observation and forecast systems began with support for offshore oil exploration at Shell Development Company (1983–1986) and moved to supporting submarine fleet operations for the Naval Oceanography Command while at Harvard University (1986–1990). At Rutgers (since 1990), he has been a principal investigator (PI) or co-PI on more than $140 million in research and application projects for the National Oceanic and Atmospheric Administration (NOAA); the National Science Foundation; the U.S. Navy; the National Academies of Sciences, Engineering, and Medicine's Gulf Research Program; and the offshore energy industry. His research interests include the development of new autonomous ocean observing technologies, their application to scientific research in remote and extreme environments, and the development of a workforce to support broader global participation in ocean observing. A major focal point has been improving our scientific understanding and our ability to operationally forecast rapid co-evolution of the ocean, atmosphere, and seabed in hurricanes and typhoons. Dr. Glenn is a fellow in the Marine Technology Society and has received the international Society for Underwater Technology's Oceanography Award as well as the Carnegie/CASE U.S. Professors of the Year award for New Jersey. He has served as the chair of the U.S. Committee for the Intergovernmental Oceanographic Commission, and is a member of the NOAA Science Advisory Board's Environmental Information Services Working Group responsible for advising NOAA on implementation of the Weather Act. Dr. Glenn is a member of the Ocean Studies Board at the National Academies.

Patrick Heimbach is a professor at The University of Texas at Austin and holds the W.A. "Tex" Moncrief, Jr., Chair III in Simulation-Based Engineering and Sciences. His research focuses on understanding the general circulation of the ocean and its role in the global climate system. As part of the "Estimating the Circulation and Climate of the Ocean" (ECCO) consortium that originated under the National Oceanographic Partnership Program, he and his group are applying formal inverse modeling techniques for the purpose of fitting a state-of-the-art general circulation model (the MITgcm) with much of the available satellite and in-situ observations to produce a best possible estimate of the time-evolving three-dimensional state over the past few decades of the global ocean and sea ice cover. ECCO products support global and regional ocean circulation and climate variability research on time scales of days to decades. Emerging research foci are understanding the dynamics of global and regional sea level change, the provision of formal uncertainties along with these estimates, and implications for improving the global ocean observing system for climate. He earned his Ph.D. in 1998 from the Max-Planck-Institute for Meteorology and the University of Hamburg, Germany. Dr. Heimbach is a member of the Ocean Studies Board at the National Academies.

Marcia Isakson is the director of the Signal and Information Sciences Laboratory at the Applied Research Laboratories at The University of Texas at Austin. Dr. Isakson received her B.S. in engineering physics and mathematics from the United States Military Academy at West Point in 1992. Upon graduation, she was awarded a Hertz Foundation Fellowship and completed a master's degree in physics from The University of Texas at Austin in 1994. Dr. Isakson served in the U.S. Army from 1994 to 1997 at Fort Hood, Texas. She earned a Ph.D. in physics from the University of Texas at Austin in 2002. She has been employed at The University of Texas at Austin since 2001 and has served as the principal investigator on more than 21 sponsored projects working with the Office of Naval Research (ONR), the Naval Oceanographic Office, and ExxonMobil. She is a key principal investigator for ONR's Task Force Ocean initiative working directly with the operational Navy. Her research interests include littoral acoustic propagation, autonomous underwater vehicles, high-frequency sonar, and high-fidelity acoustic modeling. She has extensive experience with at-sea experiments, both foreign and domestic. Dr. Isakson has taught underwater acoustics at The University of Texas since 2009. She is a former president and a fellow of the Acoustical Society of America and a Distinguished Lecturer of the Institute of Electrical and Electronics Engineers Oceanic Engineering Society. Dr. Isakson currently serves on the governing board of the American Institute of Physics and is a member of the Ocean Studies Board at the National Academies.

Lekelia (Kiki) Jenkins is an associate professor in the School for the Future of Innovation in Society at Arizona State University. As a National Science

Foundation Graduate Fellow, Dr. Jenkins received her Ph.D. from Duke University in 2006 by pioneering a new field of study into the invention and adoption of marine conservation technology. Since then, she has worked as an environmental consultant for the Natural Resources Defense Council, while also actively participating in the burgeoning field of Studies in Expertise and Experience. As an American Association for the Advancement of Science Science and Technology Policy Fellow with the National Marine Fisheries Service's Office of International Affairs, she helped implement new regulations to address bycatch and illegal, unreported, and unregulated fishing by foreign nations. Dr. Jenkins became a research associate at the University of Washington in 2009, where her research was supported by the David H. Smith Conservation Research Fellowship and the Ford Foundation Diversity Postdoctoral Fellowship. In 2011, Dr. Jenkins was hired as an assistant professor at the School of Marine and Environmental Affairs and was recently awarded the Alfred P. Sloan Research Fellowship and inducted into the Global Young Academy. Dr. Jenkins is a member of the Ocean Studies Board at the National Academies.

Sandra Knight is a senior research engineer in the Department of Civil and Environmental Engineering at the University of Maryland where she works with her colleagues in the development of water policy, disaster resilience, and flood risk management initiatives to support the Center for Disaster Resilience. Additionally, she is the founder and the president of WaterWonks LLC in Washington, DC. Her company was formed to capitalize on her extensive experience in federal disaster reduction, flood risk management, and marine transportation policies and programs, having spent more than 30 years administering these and other policies at three federal agencies. Dr. Knight finished her federal career in October 2012 as the Deputy Associate Administrator for Mitigation, Federal Emergency Management Agency, responsible for the nation's floodplain mapping, management and mitigation grants supporting the National Flood Insurance Program, environmental compliance for the agency, and oversight of the National Dam Safety Program. At the National Oceanic and Atmospheric Administration (NOAA), 2007–2009, she was responsible for the development of policies and strategies to ensure scientific excellence and improved performance of NOAA's research portfolio. Prior to that, she spent 26 years with the U.S. Army Corps of Engineers. Her last position with the U.S. Army Corps of Engineers was as the technical director for navigation research. She is a registered professional engineer in Tennessee, a Diplomate Water Resource Engineer, and a Diplomate Navigation Engineer. She is a member of the American Society of Civil Engineers, the American Meteorological Society, the Society of Women Engineers, Sigma Xi, and a fellow for PIANC. Dr. Knight received a B.S. from Memphis State University, and an M.S. from Mississippi State University. Dr. Knight holds a Ph.D. from University of Memphis. All degrees were in civil engineering.

Nancy Knowlton (NAS) is a coral reef biologist and the Sant Chair for Marine Science at the National Museum of Natural History, Smithsonian Institution (where she also served as the editor-in-chief of the Ocean Portal), and a senior scientist emerita at the Smithsonian Tropical Research Institute. She was formerly a professor at the Scripps Institution of Oceanography at the University of California, San Diego, and the founder of the Scripps Center for Marine Biodiversity and Conservation. Her areas of expertise include marine biodiversity and conservation, and evolution, behavior, and systematics of coral reef organisms. Her revolutionary studies of reef bleaching and speciation provide fundamental insights into differentiation and mutualism. Her work has revealed new, unexpected levels of diversity in the marine microbial environment. She is a member of the National Academy of Sciences. Her other honors include the Peter Benchley Award for Science in Service of Conservation (2009), the Heinz Award for contributions benefitting the environment (2011), election to the American Academy of Arts & Sciences (2013), and the Women's Aquatic Network Woman of the Year Award (2018). She received a B.A. in biology from Harvard University and her Ph.D. from the University of California, Berkeley, in zoology. Dr. Knowlton is a member of the Ocean Studies Board at the National Academies.

Anthony MacDonald is currently the director of the Urban Coast Institute at Monmouth University, West Long Branch, New Jersey. Mr. MacDonald was previously the executive director of the Coastal States Organization from 1998 to 2005. Prior to joining the Coastal States Organization, he was the special counsel and the director of environmental affairs at the American Association of Port Authorities, where he represented the International Association of Ports and Harbors at the International Maritime Organization on negotiations on the London Convention. He has also practiced law with a private firm in Washington, DC, working on environmental and legislative issues, and served as the Washington, DC, environmental legislative representative of the Mayor of the City of New York. He specializes in environment, coastal, marine, and natural resources law and policy and federal, state, and local government affairs. He earned a B.A. from Middlebury College and a J.D. from Fordham University.

Jacqueline McGlade is currently a professor of resilience and sustainable development at the University College London Institute for Global Prosperity and faculty of engineering; the Frank Jackson Professor of the Environment, Gresham College; and a professor, Institute for Public Policy and Management, Strathmore University Business School, Kenya. She was the executive director of the European Environment Agency from 2003 to 2013, where she was on leave from her post as a professor of environmental informatics at the University College London. Between 2014 and 2017 she was the chief scientist and the director of the Science Division of the United Nations (UN) Environment Programme based in Nairobi. Over the past 40 years, Professor McGlade has worked at the interface of sustainable development, science, society, and policy. She has established and led science and research initiatives across the United Nations (UN) and the European Union (EU) and around the world. Professor McGlade is known for her research on data and informatics and the use of earth observation, and assessments on biodiversity, climate change, natural capital accounting and ecosystems, oceans, social dynamics and Indigenous knowledge, and sustainable development. She is an author of more than 200 publications, including as the lead author and the editor-in-chief of more than 45 major EU and UN publications and research reports. She has been awarded a number of prizes and awards for her research. Professor McGlade completed her B.Sc. in marine biology, biochemistry, and soil science at Bangor University, United Kingdom, in 1977. She obtained her Ph.D. in 1980 on aquatic sciences and zoology from the University of Guelph in Canada. From 1987 to 1989, she was the Adrien Fellow at Darwin College and obtained an M.A. from the University of Cambridge. She holds honorary degrees from the University of Bangor, Keele and Kent.

Thomas J. Miller is a professor of fisheries and population dynamics and the director of the Chesapeake Biological Laboratory at the University of Maryland Center for Environmental Science. Dr. Miller's research interests include recruitment and population dynamics of aquatic animals, fish early life history, and blue crabs. His relevant National Academies of Sciences, Engineering, and Medicine service includes membership on the Committee on Sustainable Water and Environmental Management in the California Bay-Delta, the Panel to Review California Draft Bay Delta Conservation Plan, and the Committee on the Review of the Marine Recreational Fisheries Information Program. He is also currently serving as a member of the National Academies' Research Associateship Program's Panel on Life Sciences. Dr. Miller received his B.Sc. in human and environmental biology at the University of York, United Kingdom. He later received his M.S. at North Carolina State University in ecology, and his Ph.D. in zoology, also from North Carolina State University. He undertook postdoctoral training at McGill University. Dr. Miller is a member of the Ocean Studies Board at the National Academies.

S. Bradley Moran is the dean of the College of Fisheries and Ocean Sciences and a professor of oceanography at the University of Alaska Fairbanks. Prior to his appointment as the dean, he served as the acting director of the Obama administration's National Ocean Council, the assistant director for Ocean Sciences in the White House Office of Science and Technology Policy, and the program director in the Chemical Oceanography Program at the National Science Foundation. He focused on implementing federal ocean science policy and facilitating interagency efforts and partnerships on a broad range of ocean policy, resource, economic, and national security matters. Dr. Moran's principal research interests include applying uranium-series and artificial radionuclides as tracers of marine geochemical processes, and fostering economic development partnerships in energy and environmental research, technology, policy, and education. In 2007, he envisioned and implemented the nation's first Masters of Business Administration-Masters of Oceanography dual degree, the "Blue MBA." He is currently an editor for the *Journal of Geophysical Research-Oceans* and an editorial board member of the *Journal of Marine Science and Engineering* and the *Journal of Marine Research*. Dr. Moran is a member of the Board of Trustees of the Consortium for Ocean Leadership, the Board of Directors of the Alaska Ocean Observing System, the Board of Directors of the North Pacific Research Board, and the Board of the Alaska Sea Life Center. Dr. Moran earned a B.Sc. in chemistry from Concordia University and a Ph.D. in oceanography from Dalhousie University, and he conducted postdoctoral research at the Woods Hole Oceanographic Institution. Dr. Moran is a member of the Ocean Studies Board at the National Academies.

Ruth M. Perry is a marine scientist and a regulatory policy specialist responsible for offshore marine environmental regulations and policy for the Shell Exploration and Production Company. She integrates marine science and ocean technology into regulatory policy advocacy and decision-making in the areas of marine sound, marine spatial planning, ocean observing, and marine mammal and life science, primarily in the Gulf of Mexico. Dr. Perry is also responsible for helping Shell to develop public–private science collaborations, such as real-time monitoring programs with autonomous technology, to improve industry's knowledge of the offshore marine environment. Dr. Perry has more than 10 years of ocean technology research and system implementation, field experience, and ocean policy analysis, including research cruises to monitor coastal hypoxia and the offshore physical environment, deploying and operating ocean observing systems, and marine mammal observing in the Gulf of Mexico, Ecuador, and the Galapagos Islands. She is a member of the National Academies of Sciences, Engineering, and Medicine's Ocean Studies Board and Gulf Research Program Loop Current Committee, a Board Member of Gulf of Mexico Coastal Ocean Observing System, and on the Science Advisory Board for Texas OneGulf RESTORE Center of Excellence. She was recently awarded the Marine Technology Society's Young Professional Award for 2017. She earned a doctorate in oceanography from Texas A&M University in 2013.

James Sanchirico is a professor of natural resource economics in the Department of Environmental Science and Policy and an associate director of the Coastal and Marine Science Institute at the University of California, Davis (UC Davis). His main research interests are the economic analysis of policy design, implementation, and evaluation for marine and terrestrial species conservation, and the development of economic-ecological models for forecasting the effects of resource management policies. He received the Rosenstiel Award for Oceanographic Sciences in 2012, which honors scientists who, in the past decade, have made significant and growing impacts in their field; the UC Davis Distinguished Scholarly Public Service Award in 2014; and the 2010 and 2000 Quality of Research Discovery Award from the Agricultural and Applied Economics Association. He is currently the co-editor at the *Journal of the Association of Environmental and Resource Economists* and the principal investigator on the National Science Foundation–funded Sustainable Oceans National Research Training program at UC Davis. He served on the Lenfest Fishery Ecosystem Task Force; a National Academies of Sciences, Engineering, and Medicine committee evaluating the effectiveness of stock rebuilding plans of the 2006 Fishery Conservation and Management Reauthorization; a National Academies committee to review the U.S. Ocean Acidification Research Plan; a National Academies committee to review the Joint Subcommittee on Ocean Science and Technology Research Priorities Plan; and 6 years on the National Oceanic and Atmospheric Administration's Science Advisory Board (including service on the Ecosystem Science and Management Working Group and Social Science Working Group). Dr. Sanchirico received a B.A. in economics and mathematics from Boston University and a Ph.D. in agricultural and resource economics from UC Davis. He is a member of the Ocean Studies Board at the National Academies.

Mark J. Spalding is the president of The Ocean Foundation and an authority on international ocean policy and law. He is the former director of the Environmental Law and Civil Society Program, and the editor of the *Journal of Environment and Development* at the Graduate School of International Relations & Pacific Studies (IR/PS), University of California, San Diego (UCSD). Mr. Spalding has also taught at the Scripps Institution of Oceanography, UCSD's Muir College, UC Berkeley's Goldman School of Public Policy, and UCSD's School of Law. He was a research fellow at UCSD's Center for U.S.-Mexican Studies, a Sustainability Institute—Donella Meadows Leadership Fellow, and a SeaWeb Senior Fellow. He is the chair emeritus of the National Board of Directors of the Surfrider Foundation, and was the chair of the environmental law section of the California State Bar Association. He holds a B.A. in history with Honors from Claremont McKenna College, a J.D. from Loyola Law School, and a master's degree in Pacific international affairs from IR/PS. Mr. Spalding is a member of the Ocean Studies Board at the National Academies.

Lynne D. Talley is a professor of oceanography at the Scripps Institution of Oceanography at the University of California, San Diego. Dr. Talley's expertise and research interests include general ocean circulation, hydrography, theory of wind-driven circulation, and ocean modeling. She is the co-principal investigator for the U.S. GO-SHIP program; GO-SHIP is an international repeated decadal survey of the deep ocean's physical and chemical properties. She is the head of the observational team in the Southern Ocean Carbon and Climate Observations and Modeling program that is deploying a network of biogeochemical Argo profiling floats throughout the Southern Ocean. She was a lead author on the Intergovernmental Panel on Climate Change 4th and 5th Assessment Reports chapter on Ocean Observations, and on the 5th Assessment Report's Technical Summary and Summary for Policymakers. Dr. Talley has an extensive National Research Council committee background, having served previously on the Climate Research Committee; Global-Ocean-Atmosphere-Land System Panel; Panel to Review the Jet Propulsion Laboratory Distributed Active Archive Center; and committees on Abrupt Climate Change, Climate Change Feedbacks, and Future Science Opportunities in the Antarctic and Southern Ocean. She is a member of the CLIVAR Southern Ocean Region Panel and the U.S. CLIVAR Southern Ocean Working Group. Dr. Talley was a National Science Foundation Presidential Young Investigator in 1987. Dr. Talley received her Ph.D. in physical oceanography from the Woods Hole Oceanographic Institution/Massachusetts Institute of Technology Joint Program in Oceanography. She is a fellow of the American Academy of Arts & Sciences, the American Geophysical Union, the American Meteorological Society, and the Oceanography Society.

Robert (Bob) S. Winokur has more than 55 years of experience in marine science and satellite remote sensing and retired as a long-time senior executive after 47 years of federal service. As a senior executive in the federal government and private sector his positions included Deputy Oceanographer of the Navy and Deputy/Technical Director for Oceanography, Space and Maritime Domain Awareness, Office of the Chief of Naval Operations; acting Oceanographer of the Navy; Assistant Administrator for Satellite and Information Services and Acting Director, National Weather Service, National Oceanic and Atmospheric Administration (NOAA); senior executive positions in the Office of the Assistant Secretary of the Navy and the Office of Naval Research; president, Earth Satellite Corporation; and vice president, Consortium for Oceanographic Research and Education. His technical experience focused on undersea warfare, ocean policy, and satellite remote sensing. He is currently a senior advisor for the Michigan Tech Research Institute and NOAA Satellite Service, consulting on ocean and space policy and programs. Mr. Winokur has a bachelor's degree from Rensselaer Polytechnic Institute and a master's degree from American University. He has served on numerous government advisory boards and national and international committees on undersea technology, remote sensing, environmental satellite systems, ocean policy, and oceanographic ship management and planning. He is a fellow of the Marine Technology Society, the Acoustical Society of America, and the American Meteorological Society and serves as a member of the Ocean Studies Board at the National Academies.

Grace C. Young is a senior research engineer and the lead scientist at X, Alphabet's Moonshot Factory (formerly called "GoogleX"), where her team is creating radical new technology to protect the ocean while feeding humanity sustainably. An avid sailor, diver, and National Geographic Explorer, Dr. Young is passionate about developing tools to better understand, explore, and manage the ocean. She earned her B.Sc. in mechanical and ocean engineering from the Massachusetts Institute of Technology (MIT) (2014) and her Ph.D. from the University of Oxford (2018) as a Marshal Scholar. She has developed robots, imaging systems, and other technologies for MIT, CERN, the National Aeronautics and Space Administration, the Woods Hole Oceanographic Institution, and the National Oceanic and Atmospheric Administration. She assists National Geographic in various initiatives to educate and inspire young people about the ocean, including augmented reality exhibitions, games, and classroom outreach. In 2014, she lived underwater for 15 days as a mission scientist on Fabien Cousteau's Mission 31, the youngest Aquarius aquanaut at the time. A former ballerina, she is active in arts communities; her exhibition of ultra-high speed photography captured while living underwater was selected as "Best of Oceans at MIT 2015." Dr. Young was a 4-year varsity letterman on MIT's sailing team and sailed across the Atlantic for the nonprofit SailFuture. Dr. Young also serves as the chief scientist for the Pisces VI deep-sea research submarine.